IS AI GOOD FOR THE JEWS?

Is AI Good for the Jews?
The Canary in the Algorithm: What Starts with the Jews Never Stays There

ISBN (paperback): 978-1-968919-39-9
ISBN (ebook): 978-1-968919-40-5

ARMINLEAR

Armin Lear Press, Inc.
215 W Riverside Drive, #4362
Estes Park, CO 80517

IS AI GOOD FOR THE JEWS?

The Canary in the Algorithm
What Starts with the Jews Never Stays There

Craig R. Frank

In memory of my grandparents Bernard and Anna Klein,
my aunt Thelma Rose, and my mother Arlene Rosenfeld.
May Their Memories Forever Be a Blessing.

For my daughters, Addie, Noa, and Gali
May they draw comfort from knowing the antisemitism of their era is just a technological upgrade of the classic. We survived then, and we will survive now. Am Yisrael Chai.

Acknowledgments

I would not have been able to complete this book without the help, support, encouragement, and thoughtful suggestions of those closest to me. I am profoundly grateful to my family—Drora, Addie, Noa, and Gali—and to my grandchildren, Adam and River, who continue to inspire me.

My unending appreciation goes to Lawrence Goldstein, whose steadfast support and thoughtful advice has taught me that nothing is impossible when we try. I am deeply grateful to Ilan Sarid, Mitch Chupak, and Jeffrey Schwarz who lifted me at the moments I most needed it.

I thank Alma Flesch for her patience, insight, and editorial wisdom throughout this process.

With deepest love and gratitude to you all.

Contents

Introduction: You Don't Have to Be Jewish 1

Prologue: What is Antisemitism? 7

Part 1: How Earlier Tech Revolutions Shaped Jewish Fate 11

1 The Printing Press: When Information Became Dangerous 13
2 Trains, Ships & Telegraphs: The Modern World Arrives 21
3 Industrialization: Factories, Labor Movements & New Stereotypes 25
4 Radio and Film: Broadcasting the Best and Worst of Humanity 29

Part 2: The Internet Era: Jewish Life Goes Online 35

5 The Early Web: Hope, Community, and the First Digital Antisemites 37
6 The Dot-Com Diaspora: Reinventing Jewish Life Digitally 41
7 Conspiracy 2.0: How the Internet Recycled Old Hatreds 49

Part 3: Social Media and the Algorithmic Age 59

8 Algorithmic Psychology: Why the Internet Thinks Like a Brain on Fire 61
9 Meta: The Friendly Neighborhood Algorithm (with a dark side) 69
10 YouTube: the Unseen Giant 75
11 TikTok: The Teen Algorithm That Accidentally Turned Political 81
12 Twitter/X: Where Public Discourse Goes to Burn 85
13 Identity Wars: Influencers, Activism, & Digital Tribes 91

Part 4: October 7 and the Digital Breakdown 97

14 The First 72 Hours: Chaos, Misinformation & Narrative Collapse 99
15 The Victim-Villain Inversion: How the Jews Lost the Narrative in Record Time 107
16 Could the Inversion Have Happened If the Victims Weren't Jewish? 113
17 Digital Warfare: Propaganda, Bots & the Attention Battlefield 117
18 The Media Feedback Loop: How News and Social Platforms Reinforced Each Other 125
19 Governing the Algorithm: Regulating Platforms in the Era of Algorithmic Influence 145

Part 5: Artificial Intelligence & the Jews **163**
20 Understanding the Machine: What AI Actually Is 165
21 When Bias Meets Data: How AI Learns Hatred 169
22 Israel and AI: Innovation as Security 173
23 AI as Protector: Moderation, Monitoring, & Memory 179
24 Judaism, Ethics, & the Intelligent Machine 183
25 Virtual Jewish Life: AI Rabbis & Digital Shuls 189
26 Why Jews Keep Ending Up at the Center of Tech Revolutions 195

Part 6: Is AI Good for the Jews? **205**
27 So...Is AI Actually Good for the Jews? 207

Epilogue: A Blessing for the Algorithm **211**

Introduction

You Don't Have to Be Jewish

As a child in the Bronx, I vividly recall the advertisements for Levy's Real Jewish Rye posted throughout the Kingsbridge Road D-train subway station. Featuring photos of a diverse group of non-Jewish people—an Asian boy, a Native American man, an African American man—the posters read, "You Don't Have to be Jewish to Love Levy's Real Jewish Rye." That's how I feel about this book. You don't have to be Jewish to understand that what is bad for the Jews eventually becomes bad for everyone. The way AI affects the Jews will be the way AI affects us all.

Much of the attention being paid to the dangers of AI focus on the economic consequences (unemployment), a fear of autonomous machines taking over, or the use of AI by governments to surveil their populations. The war in Iran has also raised some fears that the integration of AI into weapon systems will one day be used to abdicate responsibility for war crimes. Certainly, human beings should not surrender moral agency to machines.

This same principle should apply beyond the battlefield. There must also be meaningful human responsibility in the information domain. Platforms cannot shrug and say that their systems merely reflect engagement. Engineers cannot pretend that recommender systems are neutral when they consistently reward outrage, simplification, tribalism, and spectacle. Political actors cannot hide behind "the discourse" when they knowingly exploit manipulated or unverified content. Influencers cannot wash their hands of responsibility after laundering lies into mainstream conversation. The media cannot surrender their mandate to the noise on social media.

If AI-generated deception helps drive hatred against Jews, then this is not just a moderation problem. It is a moral and civic problem. It is also, increasingly, a communal security problem because once falsehoods are repeated often enough, they do not remain abstract. They lead to threats outside synagogues, harassment in schools, intimidation on campuses, vandalism, exclusion in professional spaces, and a broader atmosphere in which Jews are treated as suspect, stained, or collectively guilty. The digital world does not stay digital for long.

Narrative shapes legitimacy. Legitimacy shapes pressure. Pressure shapes institutions. Institutions shape power. This affects everyone. Not just the Jews.

Is AI good for the Jews? might sound a bit dramatic, perhaps with a comic undertone. Maybe it's a bit too aggressively Jewish. But we ask this question because it is clear that AI will reshape Jewish life in wonderful, and just as often in devastating, ways, just as every other major technological advancement has done. Dramatic societal changes have never been neutral for the Jews.

This book is an attempt to trace that history with clarity, honesty, and, where possible, humor, because if you can't laugh at the absurdity of human invention, you'll cry at what people do with it.

Why Jews Always Ask the Hard Questions

There's a particular kind of question that only Jews seem to ask with a straight face. It usually begins with: *Is this good for the Jews?* Everything from weather patterns to presidential elections to new types of hummus is often evaluated through this lens. It's not paranoia but an understanding of the world that is deeply rooted in Jewish national memory. When you've lived a few thousand years with a talent for survival and a knack for attracting the wrong kind of attention, you develop instincts.

Beneath the jokes lies the deeper truth that Jewish anxiety is one of history's great predictive engines. Jews don't just react to change, we pre--grieve its consequences, model its scenarios, and argue its moral implications before most people have even downloaded the update. We have been engaging in what the modern world calls risk assessment for centuries, most often over animated, sometimes fierce discussions at the Shabbat table.

At that Shabbat table I am the uncle that speaks the loudest so my concerns for our future are certain to be heard by the children. Because when I look at the last five centuries of technological revolutions, Jewish vigilance begins to look less like neurosis and more like a well-honed survival instinct. The Jewish habit of asking the hardest question first has often been the difference between catastrophe and continuity.

A Brief Meditation: People of the Book to People of the Cloud

Jewish identity has always been tied to the records through which we store our collective memory. Calling ourselves the People of the Book was a brilliant branding move in a world where literacy was rare. The book was our homeland when our homeland was denied and became our survival kit when history took its unpredictable turns.

Now, in the 21st century, we are increasingly People of the Cloud, dispersed, yet globally connected. Today, we upload our texts, rituals, debates, humor, traumas, and hopes into a digital ether that stores everything and forgets nothing. The shift from book to cloud isn't just technological, it's civilizational. AI is emerging as the next medium for meaning and memory.

A Long History of Invention-Induced Anxiety

Each time a new technology emerges, Jews find themselves in the familiar but uncomfortable spot at the intersection of opportunity and danger. While innovation brings empowerment, progress also serves to widen the reach of hatred.

Consider the printing press, a marvel of human progress. The printing press democratized knowledge and spread literacy. To the Jews, it brought the ability to print prayer books without manually copying every letter. It also, however, made it easier to print antisemitic pamphlets, conspiracy tracts, and early prototypes of what we'd now call fake news. The first great information revolution gave us both the ability to mass print the Talmud and also the mass distribution of six centuries of blood libels.

Radio and film amplified everything from art, comedy, storytelling, political mythmaking, and hatred. They brought joy and escape to millions, offering comfort during the Depression and World War 2. They also carried Father Coughlin's antisemitic poison directly into people's homes. Film turned Jewish entertainers into global icons, but also gave the world the propaganda machines used by the Nazis with horrifying skill.

In today's technology wonderland, AI provides the ability to create content, and social media provides the instant and infinite distribution of content. Jews can't help but wonder, *what could possibly go wrong?*

Jewish History Belongs at the Center of the AI Conversation

Jewish history is essential to understanding the digital age. Jews have lived inside information wars for 3,000 years. AI is not a break from Jewish history, but merely its next chapter. Jewish memory, fragile as it may be, is fiercely preserved. It may just be one of humanity's most important guides for how AI will treat history. By asking *"Is AI good for the Jews?"* we are really asking "*Is AI good for humanity, and who will decide what that means?*"

When the internet arrived, many of us saw the great promise in a borderless world where pluralism could thrive, where the past's old hatreds might finally be crushed by the weight of endless information. Instead, it turned out that the technology resurrected, amplified, and globalized antisemitism faster than any had before it.

Social media changed everything, again. It doesn't just spread information; it actually rewards extreme emotion. The angrier people got, the more the algorithms smiled. And nothing fuels online fireworks quite like Jews, Israel, identity politics, or global conspiracies. Among all the topics that create massive fights, outrage, and viral storms, Jewish related content often sits at the very center of the algorithmic blast zone.

October 7, 2023, ruptured digital discourse in ways still unfolding. Suddenly, centuries old instincts kicked in. Jews everywhere were reminded that the same platforms that connected us also had the power to erase us, distort us, or turn the world against us.

Artificial intelligence is the biggest technological turning point since the internet. It may even be bigger. It could, in fact, become the most dramatic force for change the world has ever known. And Jews, efficiently multitasking between existential dread and enthusiastic innovation, have a front row seat.

On one hand, AI can help detect hate speech, flag antisemitic patterns, moderate extremist content, translate Jewish texts, preserve Jewish heritage, and strengthen their community. On the other hand, AI can absorb centuries of bias from its training data, regurgitate antisemitic stereotypes, fabricate lies, create deepfakes, and accelerate targeted propaganda on a scale we've never seen before.

So yes, we should ask the question, and we should ask it courageously: *Is AI good for the Jews?*

Still, this book is about more than AI. To understand our relationship with AI we need to explore how other technologies have reshaped civilization, and why Jews often find themselves disproportionately affected by innovation, sometimes as beneficiaries, sometimes as victims, often as protagonists in stories they didn't write.

This book is about how history repeats itself not as tragedy or farce, but as new inventions with old problems, and about identity, vulnerability, humor, resilience, and the strange Jewish talent for seeing both the promise and the peril of human progress.

It is the story of a people trained for millennia in interpretation, debate, paradox, memory, and moral imagination who are stepping into an era when those skills are suddenly central to the survival of truth itself. We've been engaging in this conversation since Babylonian times. Welcome to the discussion.

Prologue

What is Antisemitism?

This book does not seek to argue that the internet, or the digital universe it created, resulted in a distinct form of antisemitism. It will show, however, that the platforms that run the digital ecosystem willfully and proactively serve to broadcast, amplify and magnify antisemitism, particularly after October 7, 2023. With artificial intelligence becoming more ubiquitous, there is an immediate need to understand how technologies impact the spread of hatred, and what needs to be done to prevent AI, with its boundless potential, from making matters worse.

Before we venture into AI, algorithms, TikTok mobs, digital propaganda, and the newest ways hatred mutates online, we need to establish what exactly antisemitism is.

Antisemitism is often described as the world's oldest hatred. That's technically true, but not quite precise. Antisemitism is not just a hatred. It's a theory of the world, a flexible, shape shifting worldview that explains everything and its opposite by blaming the Jews. It is a conspiracy that survives by contradiction. Like the world's worst improv troupe, antisemites can interpret any narrative as proof of Jewish perfidy.

Most hatreds are loud and crude, but antisemitism is modular, adaptive, and self-updating. It is the original open source conspiracy theory that offers every society an outlet into which it can plug its fears, politics, frustrations, and fantasies.

To understand how digital platforms and AI interact with antisemitism, we must first understand the older operating system it's built on.

Antisemitism as a Story Told Over and Over Again

Antisemitism is not one thing but a bundle of myths that societies remix during moments of instability. At various points in history Jews have experienced religious antisemitism (they are Christ-killers or apostates), biological antisemitism (they are racially impure), economic antisemitism (they are agents of capitalism), political antisemitism (they are agents of communism), colonial antisemitism (they are agents of oppression), conspiratorial antisemitism (they are hidden puppet masters), nationalist antisemitism (they are ethnic outsiders), and the most recent, post October 7 addition, antizionism (they are white, settler colonialists and inherently violent).

Each era picks the version that best suits its anxieties. The software stays the same, just the graphic interface changes. This is why antisemitism resurges even in societies where Jews are few in number or absent entirely. It's not about Jews as people, it's about Jews as symbols.

Antisemitism Is a Theory of Blame, Not Just a Bias

Antisemitism is a comprehensive worldview that explains why the world is broken. It operates like a universal adaptor. Regardless of the true cause of the insecurity, plugging the Jews into the story lights up the narrative. Economic collapse? Jews. Pandemic? Jews. Lost war? Jews. Technological anxiety? Also, Jews. Artificial intelligence? Of course, the Jews. Didn't you hear Sam Altman is Jewish? Checkmate.

Antisemitism Doesn't Need Evidence, Just Emotion

When antisemites make accusations, they are not offering a hypothesis; they are expressing feelings masquerading as facts. In Nazi Germany, Jews were declared both subhuman and powerful enough to control global finance. Antisemitism survives precisely because it does not depend on reality. It depends only on narrative satisfaction and on offering simple answers to complex problems. And simple answers spread faster than accurate ones, especially online.

Antisemitism Is a Mirror That Reflects Every Society's Fears

Every society that has embraced antisemitism has used Jews as mirrors to project its own anxieties. The Roman Empire used Jews as symbols of stubbornness. Medieval Christians used them as symbols of theological threat. In 19th century Europe Jews were symbols of capitalism and modernity, in the 20th century they were symbols of globalization and decadence, and now in the 21st century Jews, refashioned as Zionists, are used as symbols of colonial power. Each version tells far more about the society embracing antisemitism than about the Jews themselves.

This is why antisemitism is such a revealing barometer. To track antisemitism is to track a society's health. A rise in antisemitism almost always signals deep instability and pending internal conflict.

Antizionism: A Distinct Form of Antisemitism

Antizionism often functions as an extension of antisemitism when it moves beyond criticizing particular Israeli policies and instead targets the legitimacy of Jewish collective existence in political form. In practice, that can look like a unique rights veto applied only to Jews. Many nations are criticized harshly without anyone arguing they therefore shouldn't exist, yet antizionist rhetoric frequently interprets perceived wrongs by the Israeli government as reason to deny the Jewish state's right to exist. When the frame becomes Jewish self-determination as uniquely unacceptable it echoes older antisemitic patterns that treated Jews as a people who may live at the mercy of others but should not be allowed to exercise sovereignty, agency, or equal standing among nations.

This overlap becomes clearer in the recurring tropes that appear in antizionist discourse, including demonization, conspiracy insinuations, collective guilt, and the demand that Jews publicly disavow Israel as a condition of moral acceptance. When Jews anywhere are held responsible for Israel's actions, or harassed under the banner of antizionism, the rhetoric is functioning less like political critique and more like a social permission slip for hostility toward Jews.

None of this means all criticism of Israel is antisemitic, but it does means antizionism has become an extension of antisemitism in that it universalizes blame and recycles classic anti-Jewish narratives (alongside new accusations). The masquerade is exposed when it becomes apparent that the goal is to delegitimize not a policy, but the right of the Jews to be making any policy at all. Antizionists seek to advance the idea that the Jewish people, alone among the nations, have no right to exist collectively and no right to sovereignty. Their entire movement is little more than discrimination dressed as principle.

Antizionism borrows much of its imagery from classic antisemitism. It presents Jews as global manipulators, foreign agents, in control of governments and media, the root of global chaos, and the orchestrators of global suffering. Yet it is somewhat distinct in that it also positions Jews as white settler colonialists practicing apartheid and committing genocide. The noun changes from Jew to Zionists, but the intent is consistent. It's classic antisemitism with a modern SEO strategy.

From the pre-State riots of 1929 and 1936 to the campus takeovers of 2023, antizionism functions as moral permission for violence. By declaring the Jewish state illegitimate, attacks on Jews can be framed as resistance rather than hatred, and empathy for Israelis becomes unnecessary.

There's really a very easy test to determine whether antizionism is antisemitism and that is to ask the question, *if Israel were not the Jewish state would the world have reacted to its post October 7 actions and policies as it did?* If the answer is no, *and it is*, then antizionism is about Jews, not geopolitics.

Why Antisemitism Adapts So Easily to the Digital Age

The internet supercharges three qualities antisemitism already possessed, virality, visuality, and conspiracy logic. When AI, a technology capable of generating infinite content, deepfakes, synthetic propaganda, and algorithmic bias, enters the picture it runs the risk of hyper-charging an already super-charged system.

Why We Need to Define Antisemitism Before Talking About AI

AI is the most powerful narrative engine ever built. It will shape how history is remembered, how stories are told, and how lies are spread. Antisemitism isn't a glitch in the system. *It is itself a system*, and it adapts as swiftly as technology does.

Which is why we started this book by defining the oldest hatred. We can't fix a glitch in civilization without naming the error code.

Part 1

How Earlier Tech Revolutions Shaped Jewish Fate

Chapter 1

The Printing Press: When Information Became Dangerous

Johannes Gutenberg, a German inventor, introduced Europe to movable type printing around 1440. The machine was a breakthrough that made the manufacture and distribution of written tracts cheaper, faster, and vastly more portable. Before this, producing a book required the patience of a saint and the handwriting of a calligrapher. With printing, suddenly there was no need for a monastery full of scribes. All that was needed was a machine, some ink, and a few good curses for when the letters jammed.

The printing press changed how Europeans thought, argued, spread rumors, and organized mobs. It was, in many ways, the first version of the internet, and like the internet, it inspired a glorious explosion of human creativity, as well as a catastrophic explosion of human nonsense. And yes, the Jews were right there in the middle of it.

The Original People of the Book Meets the Original Book Machine

By the late 1400s Jewish communities began experimenting with printing. The technology was a dream for them. No more fragile handwritten manuscripts fading after too many centuries in a damp cellar and no more rabbis arguing over which scribe's copy had the correct vowel points.

In 1475, the first dated Hebrew book rolled off a press in Reggio Calabria, Italy. From there, Jewish printers spread throughout Italy, the Ottoman Empire, and beyond, becoming unlikely Renaissance entrepreneurs.

Families like the Soncinos, whose printing house produced the first printed Hebrew bible, turned printing into a full-blown Jewish industry.

With printing, Jewish communities were able to standardize their prayer books, distribute calendars, circulate halachic rulings, disseminate Talmudic editions, and post ritual instructions. For a people spread across many lands this meant a new level of unity. Communities could follow the same holiday schedule in Krakow as in Cairo and the same commentary could be studied in Salonica as in Safed. This was globalization, Jewish style, long before anyone invented the term.

But Let's Not Pretend This Was All Enlightenment and Literacy

The printing press didn't just give people better access to wisdom; it also provided better access to humanity's worst instincts. Overnight, every village lunatic with half-baked opinions about Jews, witches, foreigners, or the shape of the Earth suddenly had a bullhorn.

This was an era without content moderation or fact checkers. Antisemitism, which previously had to travel from town to town on foot, like a door to door salesman, became portable, scalable, and horrifically efficient.

The First Viral Hate Content

One of the earliest beneficiaries of the new technology was the medieval conspiracy theory claiming that Jews kidnapped Christian children to use their blood for ritual purposes. Before the printing press, this blood libel spread through gossip and sermons. With printing, the defamatory claim was mass-produced in thousands of identical copies, complete with terrifying woodcut illustrations. It was the 15th-century version of what we now call going viral.

Then came Martin Luther, champion of religious reform, enemy of Catholic corruption, and, in his later years, author of some of the most grotesquely antisemitic tracts ever printed. His 1543 bestseller, *On the Jews and Their Lies*, was distributed widely, reprinted for centuries, and eventually quoted with admiration by the Nazis. Luther was the first man to definitively prove that the printing press could spread ideas both noble and monstrous, depending on who got to the machine.

Printing enabled stereotypes about Jews to become codified, fixed in standardized imagery. Woodcut artists produced the earliest mass-replicated caricatures of Jews with hooked noses and thick lips clutching bags of money alongside demonizing texts that set stereotypes which endure to this day. It was photoshop for the Middle Ages, but with more chisel work.

The Basel Talmud Burning: When Technology Collides with Fear

As Jewish printing grew, Christian authorities became increasingly alarmed at the sheer volume of Talmudic material flowing across borders. In 1553, things boiled over when Pope Julius III ordered a global ban on the Talmud. As a result, church authorities confiscated and burned Hebrew books across German and Italian territories (Venice, Rome, Bologna, Cremona, Ancona), destroying tens of thousands of volumes, including entire print runs of Talmudic tractates. The goal wasn't only to assault Jewish knowledge, but to disrupt Jewish unity.

It was a devastating loss for Jewish scholars. But it also revealed that technology could amplify Jewish vulnerability just as surely as it amplified Jewish resilience. Every time Jews found a way to use new tools to strengthen themselves, someone else found a way to turn those same tools against them.

Jewish Book Smuggling: A Noble, If Nerdy Rebellion

The moment the Church banned Hebrew books, Jews invented the first global supply chain, not powered by Amazon but by barrels, linen, and a few courageous merchants with questionable poker faces.

The most charmingly absurd smuggling method was the decoy cover. Jews would take a forbidden Hebrew book and have it rebound inside a perfectly innocent Christian looking cover. Customs officials would take one look at the cover and wave it through.

When disguising books didn't work, Jews turned to the ancient art of creative shipping logistics, a/k/a smuggling. Hebrew volumes traveled across Europe nestled inside olive oil crates, spice chests, cloth shipments, and wine barrels. So was it that 16th century customs officers were disinclined to slice open a perfectly good barrel of wine just to check if by chance hidden inside was a book of rabbinical commentary.

Some of the most reliable smugglers were, surprisingly, Christian merchants. Especially in Venice, Ferrara, and Amsterdam, Christian traders quietly shuttled Hebrew books across borders for Jewish customers. Archival contracts show Jewish clients paying extra for "discreet delivery," and merchants hiding Hebrew texts under piles of Catholic literature.

And then there were the book runners, literal couriers whose job was to carry small quantities of banned texts across borders in personal luggage. They walked miles through forests, bribed guards, memorized patrol schedules, and delivered books to rabbinic leaders like clandestine operatives. Some were arrested and jailed, and others became quiet folk heroes. Their work kept Jewish learning alive in cities where entire libraries had been confiscated.

In places where smuggling was too risky, Jewish communities set up underground channels, passing one hidden copy of a banned text from hand to hand, with sections copied by candlelight. A single Talmudic volume might circulate through hundreds of families, accumulating smudges, wax drips, and whispered arguments along the way.

And then, as if persecution were not exhausting enough, came one of the strangest episodes in the history of Jewish printing, the Cremona catastrophe of 1559. In this small Italian city two rival Jewish printing houses, the Foa family (cozy with one Catholic faction) and the Riva family (cozy with another), battled for market dominance. When the Riva press issued a new edition of the Talmud, the Foa printers reported their Jewish rivals to the Inquisition, claiming the edition contained blasphemies.

The Inquisition confiscated and burned every Hebrew book in Cremona, roughly 10,000 volumes. It was one of the largest Jewish book burnings in European history, triggered not by external antisemitism alone but by internal rivalry. As a result, entire regions suddenly had no print supply, and Jews from Lombardy to Poland were begging for a copy of the books that had gone up in flames.

Amsterdam: A Printing Promised Land

With Italy and Spain increasingly hostile, Jewish printers learned to do something modern tech companies still struggle with, pivot. When Italy got too dangerous, they picked up the whole enterprise and moved, sometimes in the middle of the night, to Amsterdam, the one place in Europe where censorship had not yet crushed the spirit of Hebrew scholarship. In the 1600s, the Dutch Republic became the print capital of the Jewish world, and its presses produced Hebrew Bibles, Talmuds, prayer books, and philosophical works, all with beautiful typefaces and unusually few threats of arrest.

Amsterdam Hebrew books flowed across the Jewish world, legally to Morocco, Poland, and the Ottoman Empire, and illegally right back into Italy and Germany through the same smuggling networks as before. Italian and German Jews treasured Amsterdam editions with their crisp Hebrew letters and clean layout, and the books became symbols of spiritual liberation.

By the late 17th century, Amsterdam printers had achieved something extraordinary by turning a marginal, persecuted printing tradition into a thriving, international Hebrew book industry. And they did it with talent, audacity, and a smuggling network that would put most intelligence agencies to shame.

Jews in the Enlightenment

The explosion of Jewish printing in Amsterdam didn't just preserve tradition, it laid the groundwork for Jewish modernity. As Hebrew books flourished, Amsterdam became a magnet for early secular Jewish thinkers and philosophers. Hebrew printing houses weren't just producing prayer books anymore. Now, they were producing ideas, some dangerous, some liberating, and, occasionally some, like those of Spinoza, heretical.

If modern Jewish intellectual life has a single founding architect, it is Moses Mendelssohn, the unlikely, soft-spoken philosopher who became the spark of the Jewish Enlightenment (Haskalah). In the late 18th century, Mendelssohn insisted that Jews could be fully Jewish and fully modern at the same time, a radical idea in a Europe that preferred its Jews segregated, silent, or invisible. He translated the Hebrew Bible into elegant German, opening Jewish texts to the broader world and opening the broader world to Jewish minds. Mendelssohn argued that Judaism was a religion of reason, ethics, and intellectual freedom, not dogma, and that Jews had something vital to contribute to the new age of philosophy, science, and civil society. His project didn't merely modernize Jewish thought, it completely rewired Jewish self-perception. The Haskalah birthed a generation of Jews who embraced literacy, integration, and intellectual ambition, traits that would later propel Jews into universities, laboratories, cultural institutions, and, eventually, to the technological revolutions of the 20th and 21st centuries. In a very real sense, Mendelssohn was the first modern Jewish futurist.

The Science of Judaism

Long before Jews found themselves overrepresented in Silicon Valley think tanks and AI labs, they helped pioneer an earlier intellectual revolution known as *Wissenschaft des Judentums*, the Science of Judaism. Emerging in 19th-century Germany, its aim was to study Judaism with the same scholarly precision applied to classical civilizations, philology, anthropology, and emerging scientific disciplines.

Instead of treating Jewish tradition as a sealed, sacred inheritance, the scholars of Wissenschaft dissected it, analyzed it, and reassembled it as a dynamic cultural system. They treated Jewish texts like data sets and Jewish history like a laboratory of human meaning. This movement trained generations of Jews to approach their own heritage through rigorous and critical inquiry. It gave Jews permission to combine reverence with intellectual audacity.

In many ways, The Science of Judaism was the first modern Jewish tech revolution. It set the blueprint for how Jews would later blend tradition with innovation and memory with analysis to identify the tools of modern knowledge. It helped shape a Jewish mindset that could both honor the past and interrogate it, which is the exact cognitive posture Jews would bring into every technological frontier that followed.

All of this, the Enlightenment debates, secular Jewish philosophy, and early scientific engagement, was made possible because Jewish books existed in portable, replicable form. The printing press didn't just preserve Judaism, it expanded it, diversified it, and propelled it into the intellectual bloodstream of Europe. For the first time, Jews could participate fully in the world of ideas, not just as custodians of ancient texts, *but as creators of new ones.*

The Internet Before the Internet

It's tempting to imagine that the arrival of the internet taught us something new about human nature, but it didn't. The printing press revealed the exact same lesson half a millennium earlier, showing that technology does not democratize truth; it democratizes access. And everything—truth, lies, brilliance, and hatred—all pour through its gates at the same time.

A Jewish Conclusion to a Very Old Problem

For Jews, the printing press was a blessing and a curse. It provided a gateway to modernity yet created a new vulnerability. It preserved texts that might have been lost forever, unified communities across continents, and strengthened Jewish intellectual life in extraordinary ways. But it also gave antisemites a megaphone to spread lies faster than ever before. It made hatred reproducible, and it taught the world, long before TikTok, Facebook, or X, that the moment information becomes inexpensive and sharable, someone, somewhere, will use it to demonize the Jews.

Still, Jews embraced the printing press. They adapted, innovated, and refused to disappear, as they have always done. That's what makes this whole story of printing presses, internet platforms, and now artificial intelligence not just a history of technology, but a history of Jewish survival.

Chapter 2

Trains, Ships & Telegraphs: The Modern World Arrives

If the printing press was the first great technological shock that forced Jews to ask, "Is this good for us?" then the 19th century delivered an even bigger surprise in a transportation and communication revolution so dramatic that it made the previous centuries feel like it they were still downloading.

Suddenly, humanity wasn't crawling across continents by horse or waiting weeks for letters to arrive. Now there were trains and steamships to transport people, and the telegraph to send messages at the speed of electricity. It was the birth of modernity. Which meant it was also the birth of new ways to misunderstand, fear, exploit, and stereotype Jews.

The World Speeds Up and So Do the Jews

In the late 19th and early 20th centuries, especially between 1880 and 1924, more than two million Jews left Eastern Europe in one of the largest mass migrations in history. Escaping pogroms, poverty, and impossible restrictions, they fled toward hopefully greener pastures in North America, South Africa, Argentina, and Ottoman Palestine.

Trains carried Jews from shtetls in Ukraine and Lithuania to ports in Hamburg, Bremen, and Antwerp. Steamships ferried entire families across the ocean, often in steerage and always seasick. Telegraphs let immigrants send word home that they had arrived safely, found work, or met their future spouse.

In 1837, when Samuel Morse sent the first telegraph message, he probably expected it would be used for important matters like national security or scientific progress. Instead, within a few decades, it was being used to spread gossip, trade rumors, get stock manipulation tips, and the occasional antisemitic hoax. It was the Victorian internet.

The Birth of Jewish Urban Worlds

These technologies didn't just change Jewish destiny; they transformed Jewish geography. They turned impoverished enclaves into a global network.

By the early 1900s, the new transportation world had created entire Jewish cities within cities like New York's Lower East Side, where Yiddish theater bellowed, tenements multiplied, and newspapers shouted headlines in Hebrew block print. Or London's East End, where kosher butchers, sweatshops, anarchists, and Torah scholars lived elbow to elbow. The Jews flocked to cities because they were where the ports, factories, telegraph offices, and rail hubs were. When modernity became urban, so did the Jews.

The Telegraph: The First Jewish Internet

The telegraph was arguably the most influential invention in Jewish communal life since the introduction of the portable God concept following the Babylonian exile. Suddenly, Jews could send money to relatives across continents, request help during crises, or share news about pogroms. Jewish philanthropy graduated from front door requests for charity to organizations collecting donations via telegraph. Jewish philanthropic networks became some of the first global NGOs, powered by wires stretching from Wall Street to Warsaw.

When an antisemitic riot broke out in Kishinev or Odessa, Jewish leaders in London or New York knew within hours, not weeks. Emergency funds moved across the Atlantic with unprecedented speed. Jacob Schiff, a German-Jewish banker in New York, famously used telegraph lines to funnel massive philanthropic aid into Russia.

But the telegraph also had a bit more troubling impact. It enabled antisemitic rumors to go viral. If the printing press had helped standardize stereotypes, the telegraph helped distribute them at almost lightning speed. In the Victorian era, a rumor about Jews manipulating the grain market could travel from Berlin to Budapest to London in a single day. It was Twitter before Twitter, with roughly the same ratio of panic to fact.

Modern Antisemitism, Powered by Steam

With the rise of railroads and national markets, new societal anxieties emerged, and they came with new stereotypes. Jews, who had suddenly become mobile, urban, and visible in professions tied to money and trade, were described by nationalists as rootless cosmopolitans, wandering financiers, and foreign infiltrators. Never mind that European aristocrats had been running multinational estates for centuries, Jews getting on trains apparently symbolized global instability.

And then, in 1903, came one of the most enduring pieces of antisemitic fiction ever created *The Protocols of the Elders of Zion*. A plagiarized fever dream about a Jewish plan for world domination, the booklet wouldn't have amounted to much in the Middle Ages. But in 1903, trains shipped it across Russia, steamships carried it to Paris, London, and New York, and telegraphs enabled foreign journalists to report on it quickly.

Baron Hirsch and the Jewish Railway Project

One of the most fascinating Jewish figures of the era was Baron Maurice de Hirsch, a German Jewish philanthropist who believed the solution to European antisemitism was mass resettlement. With the zeal of a man building a utopia and the wallet of a man who could buy one, Hirsch funded railway lines in the Balkans and financed Jewish agricultural colonies in Argentina. He essentially tried to create an entire alternative Jewish future using trains. It was the 19th-century version of a Mars startup.

Migration, Technology, and the Eternal Problem

By the early 20th century, the pattern was clear; migrating Jews empowered by new technologies established new opportunities for settlement, safety, and prosperity. It also meant new accusations and more efficient, speedy ways to disseminate them.

A century later, only the rhetoric has changed, a reflection of the fears and concerns of each era's antisemites. Whereas once the Jews were capitalist, communists, and wandering, today they are Zionists occupiers, globalists, and elitist. Technology keeps forcing antisemites to change the vocabulary.

A Jewish Conclusion to a Very Modern Problem

Trains and steamships carried millions to freedom, prosperity, or at least apartments on the Lower East Side. Telegraphs turned the Jewish world into a global family, connected, responsive, mobilized.

Modernity arrived on steel tracks and copper wires. Jews climbed aboard because they had no other choice. They entered a new world with the hope, determination, and a devotion to learning that had been their survival toolkit for two millennia. But riding the modern world also meant bracing for the backlash, a cycle that would repeat again with radio, film, the internet, social media, and now AI. As technology opened the world to the Jews, not surprisingly, the world wasn't quite ready.

Chapter 3

Industrialization: Factories, Labor Movements & New Stereotypes

While the technologies of the 19th century led to the great migration of Jews across continents, by the late 19th and early 20th centuries, something even more dramatic was shaking up the world—machines that could produce shoes, shirts, sausages, and business revolutions faster than any human ever could.

Industrialization hit Europe and America like a thunderstorm made of soot, steam, and overtime. Factories rose, cities ballooned, and workers organized. National economies transformed overnight. Everything was noisy, smoky, crowded, and astonishingly fast.

And, of course, somehow Jews ended up in the center of it, just as industrial capitalism created brand new social anxieties. Anxious societies have a long, well-documented history of pointing at Jews whenever they need someone to blame.

Welcome to the Machine Age

Between 1750 and 1900, the Western world vaulted from handicraft to mass production. Machines replaced artisans and mills replaced cottages. Capital became increasingly important in financing this process, and new classes emerged. The bourgeoisie owned the factories, and the proletariat broke their backs in them.

For Jews, the Industrial Revolution offered two very different paths. Some, especially Jewish immigrants in America, became workers, and some became owners or financiers. The workers filled the garment factories of New York, Chicago, Philadelphia, and Boston, many sewing 12 hours a day in unsafe, overheated rooms like the one that would make tragic history on March 25, 1911, when 146 garment workers, mostly young immigrant women, were killed in the Triangle Shirtwaist factory fire. The owners and financiers leveraged Jewish trade networks and entrepreneurial survival skills honed over centuries. This dichotomy was a setup for disaster. Industrial society, confused by its own upheaval, soon created its own folk tale that Jews were simultaneously capitalist kings and socialist saboteurs. Which, if nothing else, shows the impressive flexibility of the antisemitic imagination.

The Rise of the Jewish Capitalist Trope

As factories multiplied, so did banks, credit institutions, stock exchanges, and insurance firms. These new financial entities needed educated, multilingual clerks and managers, which matched the profile of many Jews. In Germany, Austria, France, and Britain, Jews became visible in the new middle class. This upward mobility escorted into existence the stereotype of the Jewish banker, the international financier, the mysterious figure supposedly moving global markets.

The Rothschild family, wildly successful, deeply philanthropic, and astonishingly competent, became the unwilling mascots of this fantasy. Never mind that the Rothschilds were one wealthy family among thousands of European banks. Conspiracy theories don't need math, they need something to soothe the national ego. And nothing soothes a struggling farmer or unstable monarchy like believing that someone else, ideally a small, conspicuous minority, is secretly in charge.

Soon politicians, pamphleteers, and demagogues began peddling the idea that Jews were tying nations together with invisible threads of credit, interest, investment, and loans. It was a way to explain modernity without having to understand it, not different from watching today's TikTokers trying to explain the Middle East.

Meanwhile... Jews Were Also Organizing the Workers

Here's the twist. At the very same time that Jews were being accused of running the capitalist order, Jews were also helping organize the labor movements, trying to disrupt that order.

In Eastern Europe, the General Jewish Labor Bund became a powerhouse of socialist organizing. Founded in 1897, the Bund fought for workers' rights, Yiddish culture, and political equality. Jewish labor leaders were fiery orators, tireless activists, and occasionally a little too enthusiastic about pamphleteering (a tradition still alive on Twitter).

In America, Jewish garment workers joined unions, led strikes, and helped build the modern labor movement. The 1909 "Uprising of the 20,000", one of the largest women's labor strikes in U.S. history, was largely led by young Jewish women demanding better conditions in New York's factories.

Industrialization Creates Modern Antisemitism

Historians generally mark the late 19th century as the birth of modern political antisemitism, the kind rooted not in religion, but in economics, nationalism, and racial pseudoscience.

Industrialization was a major driver in the introduction of personal mobility, urbanization, modern finance, and industrial competition. Railroads allowed Jews to move freely between towns and nations, but with mobility came suspicion. Urbanization produced fear of replacement as Jews, like millions of migrants, poured into cities. But only Jews were accused of secretly trying to take over. Modern finance fueled conspiracy theories. As economies globalized, people panicked and panic needs a villain. Industrial competition led to scapegoating. Wherever workers feared job loss or owners feared bankruptcy, Jews became the go to reason.

Modern antisemitism didn't come from ancient religious hatred alone, it came from the anxiety of societies trying to adapt to machines faster than the human psyche could adjust.

The Dreyfus Affair: A National Nervous Breakdown

If one event symbolized the industrial era paranoia about Jews, it was the Dreyfus Affair in France. For twelve years (1894–1906), Captain Alfred

Dreyfus, a Jewish officer, stood falsely accused of espionage and sentenced to life imprisonment on Devil's Island.

The impetus was a divided, industrially unstable, nationalistically insecure France, paranoid about Germany after its defeat in the Franco-Prussian War and falling behind other European competitors. Dreyfus became a convenient scapegoat. His trial, while inspiring Zionism, also set off a national flame war long before the French had access to YouTube comments.

Industrial Antisemitism Evolves Into Global Tropes

By the early 20th century, a fully modernized antisemitic narrative had emerged that claimed Jews run finance, labor, and international trade, that Jews are capitalists running cabals and communists plotting revolutions, and that these plans are aided by the Jews who own all the newspapers.

While none of these fantasies were true, they were emotionally satisfying to societies frightened by industrial change. Just as today, where "globalists," "elites," and "shadowy influencers" are often coded language for Jews, industrial era conspiracies helped people cope with the shockwaves of modernization.

A Jewish Conclusion to a Very Industrial Problem

Industrialization was a revolution. It lifted millions out of poverty, created modern life, and moved humanity into the machine age. It also created new stereotypes that would haunt Jews for the next century and fuel the darkest chapter of European history.

But Jewish communities adapted, resisted, organized, and built new lives in the modern world. They fought for workers' rights, built global businesses, and produced artists, activists, and intellectuals who reshaped culture. They were, once again, both part of the revolution and blamed for it.

Technology created new types of antisemitism. And the Jews, as always, had to navigate between the opportunities and dangers of a world changing faster than anyone could fully understand.

Chapter 4

Radio and Film: Broadcasting the Best and Worst of Humanity

If trains and the telegraph shrank the world, radio and film brought the entire world directly to the people. Before the 20th century, news arrived by newspaper, gossip, or the guy in the market who claimed he knew a guy who knew a guy. Now, voices traveled across continents, and moving pictures flickered in darkened theaters carrying stories, fantasies, fears, and political messages straight to the masses.

For Jews, this new culture shaping industry was once again both an unprecedented opportunity and a looming catastrophe.

Hollywood: A Very Jewish Dream Factory

Let's get this out of the way: Hollywood was built almost entirely by Jewish immigrants with questionable English but excellent instincts for what Americans wanted to watch—films with big emotions, happy endings, and very dramatic lighting.

The founding fathers of American cinema included Adolph Zukor, the Hungarian born founder of Paramount, Carl Laemmle, the German-born founder of Universal, Louis B. Mayer, the Ukrainian founder of MGM, Harry and Jack Warner, the Polish brothers who founded Warner Bros., and Samuel Goldwyn, the Polish born founder of Goldwyn Pictures (eventually MGM).

These men were all Eastern European Jews who fled pogroms, arrived with little more than ambition, and somehow built the industry that would define global storytelling for the next century. They were scrappy, theatrical, hungry, and unwilling to fail.

Hollywood gave Jews something they hadn't had since the destruction of the Second Temple and the birth of Christianity, cultural influence, though still without political power. It also gave America something it had never had before, an entertainment industry where anxious newcomers, not old elites, controlled the narrative. This, of course, did not go unnoticed.

Jews and the Birth of Modern Media

Long before Hollywood moguls were inventing the American dream on celluloid, Jews were already quietly, often invisibly, disproportionately shaping the earlier stages of modern media.

Photography, for example, was intertwined with Jewish innovators and experimenters across Europe. Jewish chemists helped refine early photographic plates, and Jewish studio owners in Vienna, Budapest, and Berlin turned portrait photography into a booming business. By the late 1800s, if you needed a respectable bourgeois portrait, odds were good you'd end up in front of a camera run by someone named Rosenfeld, Goldstein, or Szekely.

Newspapers too became an important Jewish sphere of activity. Jewish printers edited some of the first major Yiddish newspapers, turning cities like New York, Warsaw, and Vilna into ecosystems of news, satire, political debate, literary criticism, and gossip. The Yiddish press became the original social media network hosting heated arguments, niche communities, serialized drama, political organizing, bad takes, and brilliant essays.

Behind that world stood armies of Jewish women who served as typesetters, editors, translators, and columnists, and whose labor was largely invisible. They set the metal type by hand, inked presses, proofread columns, and helped create the infrastructure of a modern information economy. When film arrived, these same skills and sensibilities flowed directly into Hollywood, where Jewish families already had decades of experience turning words, images, and ideas into portable culture.

Even early scientific foundations of modern media, like mathematics, optics, and physics, were marked by Jewish thinkers whose work would later enable radio waves, microphones, projectors, and film reels.

In other words, by the time Hollywood's lights came on, Jews had already spent a century building the machinery, skills, and intellectual ecosystem that made mass media possible. When radio and film arrived, Jews didn't need to join the revolution: They were already standing in the room when it began.

Radio: The World's First Mass Microphone

By the 1920s, radio had become the fastest mass communication technology in human history. It could reach millions of people instantly. That was terrific when they were broadcasting jazz, comedy, or presidential fireside chats, but less wonderful when the broadcasts spewed hate speech, conspiracy theories, or propaganda. Radio permitted a miracle of modern communication to become the instigator of occasional mass hysteria.

Still, radio was a gift to Jewish entertainers. On the American airwaves, listeners fell in love with Jack Benny, whose comedic timing was so sharp he could get a laugh with a silence, and with Eddie Cantor, whose vaudeville energy transferred beautifully to the microphone. There were dozens of lesser-known Yiddish comedians, storytellers, and musicians who also found an unlikely home in radio.

Yiddish theater companies even began recording performances, bringing the sounds of the Ashkenazi stage into the homes of immigrants desperate for cultural memory. These scratchy early recordings of comic monologues, sentimental ballads, raucous operettas, and dramatic retellings of biblical legends became lifelines for people who had left behind entire worlds.

But something else was happening at the same time. Those recordings didn't stay confined to tenements and neighborhood shops. They began to circulate, influencing vaudeville performers, Broadway composers, radio writers, and eventually Hollywood storytellers. The cadences of Yiddish humor, the rhythm of Yiddish speech, the flair of Yiddish plots all seeped into the broader American cultural bloodstream. For the first time, Jewish creativity didn't just sustain Jewish communities. It shaped global culture itself, infusing American entertainment with new comedic timing, new emotional vocabulary, and a storytelling sensibility that blended satire, sorrow, resilience, and wild theatricality. The Jewish immigrant stage became the secret scaffolding of modern popular culture.

But there was a shadow creeping behind these new technologies.

Early Radio Propaganda: Early Viral Algorithm Content

Long before Silicon Valley discovered the dark arts of *engagement optimization* early radio was the first algorithmic feed. In the 1930s, broadcasters figured out that the programs attracting the largest audiences weren't polite literary readings or educational science lectures, but fiery speeches, sensational rumors, emotional storytelling, and messages that promised to reveal a hidden truth.

The Nazis understood this instantly. So did American demagogues. They created content that hit listeners' emotional reward centers, such as fear, outrage, grievance, and belonging. Radio was *sticky*. It kept people listening because it made them feel, and the feeling was usually anger or urgency.

If you squint, 1930s radio looks a lot like today's algorithmic platforms, where the content that spreads fastest is the content that punches hardest. The difference is that in the 1930s, there was no *click to share*. While radio had virality, it was slow and arduous. People simply told their neighbors what they'd heard on the radio and the snowball rolled from there.

The Dark Side of the Airwaves: Enter European Fascism

No one understood the power of broadcast media better than Joseph Goebbels, Hitler's Minister of Propaganda, who, in a March 1933 speech, famously called radio *"the most modern and most crucial instrument that exists for influencing the masses."* Goebbels turned radio into a German offensive weapon. For the first time in history, a state could broadcast hatred, electronically dehumanize a minority, easily unify a nation around racial ideology, and coordinate mass violence, all with the twist of a dial. Germans in 1933 proved to be unable to resist radio propaganda, just as today people are unable to resist viral misinformation on Instagram.

Hitler banned Jewish voices from the airwaves, silencing Jewish musicians, journalists, and entertainers. Then he filled those same frequencies with speeches, rants, and carefully constructed lies about Jewish conspiracies, racial threats, and national purity. For millions of Germans, the sound of Hitler's voice became a daily ritual, like a podcast with only one episode, repeated endlessly.

Film Propaganda: Highly Produced Disinformation

Film is the granddaddy of the deepfake. Before television, before YouTube, and before TikTok edits, there were propaganda films like Triumph of the Will, cinematic masterpieces in terms of production technique, and moral catastrophes in terms of content.

These films used dramatic lighting, stirring music, heroic camera angles, manipulative edits, staged scenes, and actors disguised as ordinary citizens to create an alternate reality that audiences accepted as truth. These were the influencer reels of the 1930s, but with government budgets and professional cinematographers.

The goal was not entertainment but persuasion through emotion, awe, and fear. If radio was the viral tweet, film was the viral video. And the Nazis understood both better than almost anyone of their time.

Today, we worry about highly produced disinformation campaigns on social media. Back then, the technology was less advanced, but the strategy was identical—*make the lie visual and people will believe it.*

The Eternal Jew

The notorious 1940 film *The Eternal Jew* presented Jews as rats, parasites, and societal invaders, using cinematic techniques like lighting, close-ups, and dramatic music to make hatred feel like truth. It was highly polished and well produced. It was proof that technology can deliver either culture or cruelty depending on who holds the camera. While Jewish producers in America were making *The Wizard of Oz*, *Casablanca*, and *The Jazz Singer*, Nazi filmmakers were doing their own dark version of storytelling, one that helped prepare a nation for genocide.

Meanwhile in America... Suspicion Brews

Back in the U.S., Hollywood's Jewish founders faced a different challenge in the form of Middle America's paranoia. Americans loved movies but didn't always love the idea that Jewish immigrants were deciding what stories they saw. This suspicion became a recurring part of American antisemitism, claiming the Jews control Hollywood, distort our morals, and corrupt American innocence. It didn't matter that Hollywood churned out deeply American mythology like stories about cowboys, gangsters, and small town heroes, all designed to show the public that they were American first and last. People still whispered.

The irony, of course, is that the Jewish moguls created Hollywood because they weren't accepted by America's elite cultural institutions. Having been excluded, they built their own, and then they were resented for succeeding. This demonstrates the classic Jewish modernity loop whereby step 1 is to survive, step 2 is to innovate, and step 3 is to get blamed for innovating.

The American Airwaves Turn Hostile

If Hollywood had its whispered antisemitism, American radio had something much louder, Father Charles Coughlin, the antisemitic Catholic priest whose broadcast sermons reached over 30 million listeners in the 1930s. Father Coughlin was the original viral influencer, except instead of selling skincare products or meal kits, he sold conspiracy theories about Jews controlling banks, sabotaging nations, and corrupting Christian civilization. He proved that radio could be used to harness mass grievance, a model replicated often in later political talk radio and, much later, in the social media echo chambers.

A Jewish Conclusion to a Broadcast Problem

American radio empowered Jewish comedians to become household names and also empowered antisemitic preachers to blame Jews for the Great Depression. Film gave Jewish immigrants control of a storytelling medium that reshaped the world. Both enabled Nazi Germany to turn Jews into symbols of evil.

And so, as with printing presses, trains, steamships, and the telegraph, Jewish communities once again discovered with radio and film that *the machine doesn't choose its message, people do.*

Coming next were the technologies that went beyond broadcasting content by connecting every person to every other person, globally, instantly, and chaotically. The internet was on the horizon. And things were about to get very, very complicated.

Part 2

The Internet Era: Jewish Life Goes Online

Chapter 5

The Early Web: Hope, Community, and the First Digital Antisemites

The early internet, what we sometimes nostalgically call Web 1.0, was a strange, hopeful, profoundly weird place. Imagine a global library built by amateurs, run on dial-up, and filled with equal parts brilliance, nonsense, and arguments about Star Trek continuity. Now add academics, hobbyists, night owls, and the occasional conspiracy theorist, and you've got the early web.

For Jews, it was a revolution. For antisemites, it was a gift. And for everyone else, it was the dawn of humanity's collective realization that giving the entire world a keyboard might have been a mistake.

Welcome to Web 1.0: The Digital Frontier Nobody Understood

Between 1990 and 2000, the internet shifted from government experiment to public playground. Websites were static, usually ugly, and often created by people who thought Comic Sans was a good idea. Chat rooms were chaotic and the message boards were endless.

At this early stage, internet culture was dominated by academics, hobbyists, introverts, and people with strong opinions about Linux. There was exactly zero adult supervision. It was the world's first massive, unregulated conversation. It was also the first time Jews could speak to one another collectively across continents without relying on synagogues, newspapers, or the postal service.

Jewish Life Goes Online . . . Fast

Jews have always been early adopters of communication tools, and the internet was no exception. By 1992, Jewish communities were already building digital infrastructure with *Shamash*, the first major Jewish internet service with a directory, discussion platform, and resource center. It was only the start of a wave of chat rooms, virtual minyans, dating sites, Jewish LBGTQ safe spaces, and student networks. Halachic debates, philosophy discussions, and arguments about Israeli politics were all conducted through long email threads.

The First Jewish Online Dating Pioneers

Somewhere between the Talmudic flame wars on Usenet and the first cyber hate reports, another very Jewish experiment was unfolding online: digital matchmaking. JDate, launched in 1997, was more than just a website. Now, with a username, a search filter, and a refrain about *"meeting their bashert"* previously insurmountable barriers were erased, all through a 56k modem.

Early JDate profiles were masterpieces of cautious self-presentation. Men described themselves as outgoing, family-oriented professionals who enjoy long walks and Israeli dancing. Women were spiritual but not too religious, and into Shabbat dinners, indie films, and travel. The men were always funny, and most of the women were down-to-earth, equally comfortable in jeans or a little black dress.

The technology was clunky, using dial-up connections, slow-loading photos, and chat windows that froze at crucial flirtation moments, but the cultural innovation was enormous. For the first time in history, single Jews in Cleveland, Miami, Los Angeles, and small town Iowa could browse each other like a highly selective, slightly neurotic catalog. Rabbis, matchmakers and over involved friends might still be arranging introductions, but now so were algorithms.

In retrospect, early Jewish online dating was one of the most successful digital projects of Web 1.0. It strengthened communal continuity and expanded the pool of potential partners. If the early web sometimes felt like a basement full of odd obsessions, JDate was the corner where hope, romance, and user generated chemistry quietly flourished.

The internet offered what Jewish history had rarely afforded, a space for Jewish visibility without immediate danger. It lasted a few years.

The Internet's Other Welcome Mat: Early Extremism

If the early web was a miracle for Jewish connection, it was also a megaphone for exactly those people who should never be given one. *Stormfront*, started in 1995, was one of the first and most influential white supremacist websites. Crude by modern standards, but revolutionary for its time, it allowed extremists to create a sense of community, publish propaganda, coordinate actions, and recruit globally. Gone were the good old days of hate when they had to meet in basements or hand out pamphlets.

On Usenet Newsgroups, debates raged between amateur historians armed with facts and deniers armed with nothing but confidence and a disturbing amount of free time. The site *alt.revisionism* became the main arena for Holocaust deniers. The anonymity was gasoline, the borderless nature was oxygen, and the early extremists were more organized than anyone realized.

How the Early Web Accelerated Antisemitism

The first contributing factor to the polluting of the early web is anonymity, which empowered boldness. People who would never spew antisemitic slurs in public suddenly felt comfortable doing so behind usernames like "Aryan-Warrior88" or "JudeoTruthSeeker1993."

Also, the Internet provided borderless publishing, which permitted print-free, mail-free circulation through easily identified distribution networks. All that was needed was HTML pages. Suddenly the messages of hate had broadband.

Finally, the permanence of websites allowed for the archiving of hate since fringe pamphlets were stored, linked, copied, and presented online forever. A search about the Holocaust could lead to denial pages because the early web had no guardrails and absolutely no algorithmic ethics. Geography was no longer a barrier, giving recruitment a global reach. A neo-Nazi in Idaho could now recruit a sympathizer in London or Sydney in minutes. *This was the moment when antisemitism transitioned from local prejudice to globalized menace.*

The First Digital Pushback

Luckily, Jews have a talent for showing up early to technological innovations and even earlier to the battles that follow.

The Nizkor Project, founded in 1991, was one of the first organized digital efforts to debunk Holocaust denial. A volunteer run project dedicated to archiving survivor testimony, providing source material, and countering misinformation on Usenet, it was Wikipedia before Wikipedia, except that every editor was uncompromising and highly caffeinated.

By the late 90s, the Anti-Defamation League built one of the first cyber hate monitoring units. Its annual reports looked like something out of a dystopian sci-fi film, but all the threats were real.

The Early Web: A Digital Basement of Obsessions

Looking back, Web 1.0 had the energy of a massive digital basement filled with amateur scholars, lonely geniuses, obsessive hobbyists, earnest seekers, and antisemitic weirdos who had finally found one another. It wasn't the scale that made it dangerous; Web 1.0 was small compared to today's Internet, but rather the precedent it established. Much like early printing presses, this was fringe, niche, and text-heavy. The groundwork was being laid.

The world didn't realize that these tiny pockets of digital hatred would become the foundation for the massive online ecosystems of the 2000s, 2010s, and 2020's. In hindsight, Web 1.0 was the warmup act, the testing ground for everything that followed. It was the first global Jewish online community and also the first global antisemitic network, two forces growing in parallel, on the same platform, with no moderators in sight. It was the first digital battlefield.

A Jewish Conclusion to a Very Early Digital Problem

The early internet gave Jews community, connection, visibility, a creative outlet, and safety (mostly). But it also gave antisemites a global stage, a permanent archive, and the ability to organize without leaving their houses.

If the printing press, the telegraph, and the radio changed Jewish life, the internet would transform it completely. As before, Jews embraced the opportunities and prepared, knowingly or not, for the coming storm.

The early web was just the beginning. The real chaos would soon arrive with social media, a technological leap that would amplify everything, everywhere, all at once. And Jews would once again find themselves at the center.

Chapter 6

The Dot-Com Diaspora: Reinventing Jewish Life Digitally

When the clock struck Y2K and the world discovered that computers were not, in fact, going to explode, humanity breathed a collective sigh of relief and promptly spent the next decade building blogs, forums, and websites dedicated to every interest imaginable, including a surprising number dedicated to ranking the greatest Seinfeld episodes.

The 2000s were the Wild West of the digital age. There were no TikTok or billion user platforms. There were no algorithms. Just a massive explosion of blogs, forums, early digital media, and comment sections where civilization went to die. For Jews, it was a golden age of digital creativity.

Welcome to the Age of Blogging, Where Everyone Became a Columnist

Between 2000 and 2010, platforms like *Blogger, LiveJournal, WordPress*, and, later, *Reddit* created a new species of humans, the online commentators who felt morally obligated to share their opinions with the universe. Naturally, Jews excelled.

Jewish digital media expanded dramatically. *Jewschool* was the scrappy, progressive Jewish blog and community hybrid, *Tablet Magazine* was intellectual, cultural, stylish, and occasionally spicy, and *Haaretz English* enabled diaspora Jews to debate Israeli politics in real time.

Suddenly, every Jewish niche had a digital home whether it was halachic debates, Zionist (and antizionist) philosophy, Jewish feminism, denominational politics, Jewish humor, Jewish cooking, or Jewish everything else. It was intoxicating and it was empowering. It was also, in hindsight, a foreshadowing of chaos.

Chabad.org: The First Truly Global Jewish Digital Institution

Long before most synagogues had a functioning website, and while many rabbis were still treating email as a suspicious fad, Chabad had already decided that Judaism needed a homepage.

Chabad.org began experimenting with the web in the mid-1990s, when going online still involved screeching modems and patient spouses. While other organizations were debating whether the internet was good or bad for Judaism, Chabad members simply assumed it was a new continent and started building shuls on it. They put up articles on every holiday, every mitzvah, and every lifecycle event, along with Ask the Rabbi forums that effectively turned rabbis into 24/7 tech support for the human soul.

What made *Chabad.org* unique was its philosophy, rather than its size. The site wasn't designed only for affiliated, observant Jews, but was built also for the curious, the confused, and the disconnected.

Over time, *Chabad.org* evolved into the first truly global Jewish digital institution. A Jew in Bangkok, Buenos Aires, or Brooklyn could get candle lighting times, read a d'var Torah, watch a class, send a question to a rabbi, and download a how-to guide for putting up a mezuzah, all from the same site. It was brick and mortar outreach translated into HTML.

In a way, *Chabad.org* was a dress rehearsal for the entire Jewish digital age, a mix of old texts and new tools, ancient law and modern technology, all held together by the quiet assumption that wherever Jews go, Judaism should already be there waiting for them.

The Internet Becomes the New Jewish Public Square

Judaism, a tradition built on debate and commentary, found its natural environment online. The early 2000s created the first truly global Jewish public square, where anyone with a Wi-Fi signal could join the conversation. You didn't need a pulpit, a professor, or a publishing contract. All you needed was a keyboard and an opinion.

Debates exploded. *Is the halachic process too rigid or too flexible? Can feminism and Orthodoxy coexist? Is Jewish Renewal brilliant or questionable or both? Is Zionism flourishing, collapsing or evolving?* It was exhilarating, and it was exhausting,

In many ways, popular Jewish practice went digital with online Torah study, downloadable Haggadot, YouTube rabbis, remote Jewish education, and early livestreams of services. The internet became an extension of Jewish life, a diaspora without borders.

Niche Jewish Communities Bloom

One of the defining features of Web 2.0 was the rise of small communities thriving because geography no longer mattered. Suddenly, there were digital spaces for Persian Jews, blogs by ex-Orthodox women, Baal Teshuva forums, Yemenite Jewish recipe groups, and academic subfields of subfields of subfields. There was even a digital home for Jewish metalheads (yes, really). For the first time, every Jewish micro community had a home. If you were a queer Sephardi sci-fi lover with strong feelings about hummus texture, you finally had people to talk to.

But While Jews Built Community, Hate Went Mainstream

While the early web had fringe antisemitism, Web 2.0 had mainstream antisemitism. Digital newspapers provided comment sections and instantly regretted it, since they opened up a new circle of hell. Articles about Israel, banking, literally anything at all, opened the door to trolls. With screen names like "UcantTellMeWhat2Think," *anonymous* users spewed antisemitic jokes, classic tropes, and conspiratorial accusations.

That was the beginning of the modern harassment ecosystem. Relieved of the need to write manifestos, online deniers—many virtually illiterate—-could now simply paste the same nonsense into any forum or comment thread. It spread like mold. Long before Pepe or QAnon, the 2000s saw early antisemitic meme templates claiming that Jews run global finance, control the media, and are the leaders of the despised global elites. There were trends claiming the Rothschilds own the weather (since reduced to merely controlling the weather).

Five Dancing Israelis

After the terror attack against the World Trade Center and the Pentagon on September 11, 2001, false accusations immediately appeared claiming the attack was the work of the Israeli Mossad as proven by the (false) "fact" that Jews knew not to show up to work that day.

On September 11, 2001, several New Jersey residents reported seeing five men filming the burning Twin Towers while appearing to be celebrating. Naturally, the police investigated and found that the men were employees of a moving company owned by Israeli nationals. They were detained by the FBI, questioned, and eventually deported for visa violations. They were not agents or operatives and were not involved in the attacks. There was zero evidence linking them to terrorism. The FBI report, released years later, found no connection to 9/11 or any hostile activity.

But, because the men were Israeli, and because conspiracy culture loves a good coincidence, their detention was exaggerated into "*agents caught celebrating,*" sparking the rumor that Mossad agents knew about 9/11. This mutated into the story that Israel planned 9/11, which plugged neatly into preexisting antisemitic tropes about Jewish and Israeli global manipulation. The story spread quickly because it fit anti-Israel sentiment, antisemitic conspiracy patterns, and was simple enough to memeify.

"Five dancing Israelis" is significant because it was one of the earliest viral antisemitic memes of the internet era, and it showed how fast conspiracies could mutate online. It revealed how Jewishness could be inserted into unrelated global events through speculation and digital rumor and was the precursor to the modern algorithmic conspiracy ecosystem. It was an early, devastating example of internet era antisemitism shaping public narratives.

The fact that it took conspiracy theorists about 48 hours to decide that five Israelis spotted in New Jersey with a camera must mean Israel masterminded 9/11 remains one of conspiracy culture's greatest leaps of illogic, representing an Olympic-level feat of mental gymnastics fueled entirely by antisemitism and dial up internet. The danger it exposed was that lies, distortions and half-truths travel faster than the truth, a pattern that would define the next twenty years.

The Iranian Holocaust Cartoon Contest

In 2006, Iran held a contest soliciting cartoons mocking or denying the Holocaust. Hundreds of entries went viral. It was one of the first times antisemitic memes crossed borders instantly, a preview of what would happen on Facebook, Twitter, and TikTok.

This contest marked the birth of the globalized antisemitic meme culture. Before 2006, most antisemitic content online was text heavy and forum based, coming from fringe corners and spreading slowly. The Iranian Holocaust cartoon contest changed that. Suddenly, antisemitic ideas were packaged in bold images, quick punchlines, emotionally charged visuals, and meme friendly, easy to share formats.

This was the prototype of social media antisemitism with content that was designed to spread fast, that required no literacy or context, and that provoked maximum emotional response. This was the earliest example of what would later become Instagram infographics demonizing Israel and viral TikTok clips spreading lies about Palestinian famine and Israeli genocide. It presented hate in cartoon form on 4chan, 8kun, and Twitter (now X), launching an era of visual disinformation campaigns, and foreshadowing the meme-driven antisemitism that dominates digital platforms today.

The Iranian contest was a geopolitical stunt engineered for virality. The contest was orchestrated by Iranian state media outlets under direct supervision of regime officials. Its purpose was to mock the Holocaust, delegitimize Jewish memory, provoke the West, and rally global anti-Israel sentiment. This use of the internet as a weapon of psychological warfare, long preceded similar campaigns by Russia, China, and extremist groups demonstrating that the internet, in addition to amplifying antisemitism, could be used as a state tool to promote hate generally. It also showed that bad actors could exploit technology platforms long before regulators would notice.

The Iranian contest also showed how the internet could aid in the shift from local to global antisemitism and coordinated online hate campaigns targeting Israel and the Jews.

The 2006 Iranian sponsored antisemitic cartoons were the moment when antisemitism became instantly exportable through digital imagery and was a direct precursor to the Gaza meme war, Holocaust jokes on TikTok, and antisemitic so-called edgy humor on Reddit.

The Iranians showed that antisemitic content thrives when it's visual, simplified, and emotional. Cartoons are powerful because they bypass intellectual analysis, stick in memory, and can be humorous, even when evil. They are easy to share and fit perfectly into the attention economy. This is the same mechanism that makes TikTok antisemitic jokes so viral, Instagram infographics so persuasive, and political propaganda memes so sticky. The contest revealed that image based antisemitism spreads further and faster because it's optimized for the human brain's visual fast lane. This was a glimpse of the future, the rise of meme-based radicalization.

The contest also exposed how online antisemitism was adapting to the internet faster than Jewish institutions. In 2006, Jewish groups were still focused on monitoring websites and forums. They were not prepared for meme warfare. They underestimated the power of visuals and had no tools for rapid response digital counter narratives. The same would be true in 2023.

The Jews were slow in their response to algorithmic hate, struggled to fight deepfakes and propaganda, and had difficulty countering rapid fire misinformation. These challenges are still being confronted today with only minimal success. While Jewish communities often embrace new technologies quickly, antisemites often weaponize them faster.

Iran is still a major driver of digital antisemitism. The 2006 contest wasn't a fluke, but the opening act of a long term strategy. Iranian state media today spread antisemitic disinformation online, supports anti-Israel meme factories, and uses bots to amplify anti-Zionist narratives and influence TikTok and X discourse. The cartoon contest is the moment the world should have realized that digital antisemitism is not random but often state crafted, state funded, and state amplified.

The Iranian Holocaust Cartoon contest marks the dawn of global antisemitic meme warfare and the shift to image based digital hate. It's the moment when antisemitic content stopped being slow, text heavy, fringe, and local, and became fast, funny (in a dark way), viral, and global.

The Wikipedia Edit Wars

No story captures the chaos of the Dot-Com Era better than the Wikipedia edit wars, a saga demonstrating that once the internet allowed anyone to write history, people who shouldn't be doing it promptly tried.

Any article involving Judaism, Zionism, Israel, antisemitism, or Middle Eastern history became a full contact sport. Anonymous editors accused each other of bias, propaganda, whitewashing, or, occasionally, simply being idiots. Entire narratives swung back and forth in real time as competing users dueled through footnotes, citations, and increasingly passive aggressive edit summaries.

These battles rather than being petty squabbles revealed something foundational about the digital age—*truth itself had become editable*. For the first time, Jewish history wasn't primarily written by scholars, journalists, rabbis, or institutions, but by whoever refreshed fastest and cited the most persuasive PDF. The familiar, structured pathways of Jewish knowledge, like books, classrooms, synagogues, and newspapers, suddenly had competition from a supposedly crowdsourced consensus, shaped by the loudest, angriest, or most persistent participants.

The edit wars showed how Jewish topics, more than almost any others, attract global ideological obsession. They also marked the beginning of a new kind of conflict, informational warfare waged through hyperlinks.

Everything that now happens on Twitter/X, Threads, and TikTok comment storms, and Reddit pile-ons started here, in these early digital skirmishes. It was the first sign that the internet wouldn't just host Jewish conversation, it would host the fight over who gets to define Jewish reality.

Most importantly, Wikipedia foreshadowed an AI-driven world where truth is even more unstable, more manipulable, and more vulnerable to distortion. The edit wars were the opening act in a larger story about how technology decentralizes authority and complicates Jewish identity. They were a digital cold war over facts, and a preview of the information chaos still to come.

How These Years Built the Architecture of Today's Internet

Everything we associate with modern digital life, and modern digital antisemitism was born during this period, including online debate culture, trolling, anonymous harassment, viral pseudoscience, conspiracy incubators, niche identity communities, partisan digital media, and extremist recruitment pipelines. Web 2.0 was proto-Facebook, proto-Twitter, proto-Reddit as we know them, laying the groundwork for the platforms that would later reshape global discourse.

A Jewish Conclusion to a Very Dot-Com Problem

The 2000s transformed Jewish identity and Jewish community. Suddenly, Jews everywhere could study, socialize, debate, create, and even organize, surmounting geographical obstacles with unprecedented freedom.

But the era also saw antisemitism scale up in new ways, moving from fringe corners to mainstream discussions, from slow spread to copy-paste virality. The internet became both the new Jewish diaspora and the new global antisemitic network, parallel worlds growing in the same digital soil.

The next decade would see those worlds collide on platforms designed not for community, but for maximizing attention, emotion, conflict, and profit. Social media were coming and everything was about to ignite.

Chapter 7

Conspiracy 2.0: How the Internet Recycled Old Hatreds

If the early web gave antisemitism a keyboard, the social media era gave it a megaphone, a marketing budget, and an algorithmic chauffeur. Between 2010 and 2020, platforms like Facebook, Twitter, YouTube, and Instagram did something extraordinary by turning conspiracy theories into premium content, meticulously tailored to grab attention, trigger outrage, and keep users scrolling long enough to see another advertisement.

Humans have been searching for whomever secretly controls the universe since we learned how to stack stones. Social media industrialized the process making conspiracies fun, visual, communal, monetizable, and available in thousands of flavors, from rage to irony to lifestyle branding. And as always, once the conspiracy machine revved up, the Jews—the historically unlucky Jews—found themselves back in the center of the storm.

The 2010s: The Age of Algorithmically Enhanced Paranoia

Social media platforms set out to maximize engagement, not super charge antisemitism. Unfortunately, nothing produces engagement quite like fear, anger, and content claiming to reveal secret cabals. Conspiracy theories are the fast food of the internet, cheap, addictive, and always available. Several major waves defined the decade.

Antisemitism Early Internet Cafes

Long before TikTok served up geopolitics between dance trends and outrageous challenges, young Jews were learning a quieter, more pixelated lesson about what happens when the world goes online. It all started with AOL.

For a certain generation, Jewish digital life began in dimly lit computer rooms and internet cafés that smelled vaguely of dust and pizza. A user logged into AOL, with that unforgettable modem screech, picked a username that felt clever (but in retrospect is somewhat embarrassing), and wandered into rooms with names like "Teen Chat," "Debate Corner," or "World Talk."

Sooner or later, that user might type that they were Jewish and, abracadabra, the energy in the room changed. Suddenly, someone dropped a slur or made a Holocaust joke. For the Jewish kid, this was often the first encounter with antisemitism without context and without an adult around to explain or intervene. There were just Jewish kids, their screens, and a flood of hostility coming from strangers with anime avatars.

The same thing happened in early online games. In RuneScape, Habbo Hotel, MapleStory, and a dozen other pixelated universes, kids from around the world chopped virtual wood, fought cartoon monsters, traded digital swords, and called each other Jew as an insult.

These early experiences are important because they formed the emotional backdrop for the era of full-blown antisemitic conspiracy culture that followed. By the time social media entered their peak in the 2010s, an entire generation of Jews had already learned that the internet was not a neutral space. It was a place where jokes about ovens appeared in game chat, where "Jew" was shorthand for greedy, and where defending yourself meant either logging off or typing back with shaking hands.

When we talk about the algorithmic storms of the 2010s, we're talking about a weather system that Jewish teens had already felt in their bones a decade earlier, back when the worst thing you were supposed to encounter online was lag.

Occupy Wall Street

In 2011, the Occupy Wall Street movement burst onto the scene with righteous anger, clever slogans, and a very understandable critique of predatory financial institutions. But as the movement migrated from physical encampments to the internet, particularly Facebook, Tumblr, and the meme heavy corners of early Twitter, its message mutated.

The 1%, a phrase meant to describe structural inequality and unchecked corporate power, became a kind of Rorschach test for the anxieties of the online masses. For most activists, it meant hedge funds, corporate greed, and political dysfunction. But in certain darker corners of the web, the financial critique slid effortlessly into the far uglier tropes about shadowy bankers, secret cabals, and global puppet masters who happened, once again, to be Jewish.

This was Victorian era conspiracy rhetoric dusted off and uploaded into the digital bloodstream. Anti-Rothschild mythology, once confined to pamphlets, fringe newsletters, and the occasional tinfoil-hatted uncle, found new life in viral graphics, Tumblr posts, and image macros that looked edgy and subversive but carried the exact DNA of 19th century antisemitic propaganda.

Globalists became the updated, socially acceptable euphemism for Jewish influencers. Charts and infographics mapped supposed networks of power that all, mysteriously, pointed back to the same surnames. What had begun as a populist movement against economic inequality inadvertently opened the door for classic antisemitic narratives to be repackaged as political critique. In the algorithmic petri dish of early social media, the old hatreds thrived.

The Sandy Hook Massacre

In December 2012, after the massacre at Sandy Hook Elementary School left twenty first graders and six educators dead, the nation entered a period of collective mourning and reflection.

Almost immediately—faster than newspapers could print the obituaries—*disinformation entrepreneurs*, a new and extremely disturbing force, emerged. These self-styled supposed truth seekers, broadcasting on YouTube, fringe blogs, and early Facebook groups, insisted the shooting was staged. They accused shattered parents of being crisis actors, hired performers working for some sinister government plot. It was an act of cruelty so breathtaking that even seasoned observers of internet culture were stunned. For the first time, the world watched as a modern mass tragedy was deconstructed, denied, and reassembled into a conspiracy theory in real time. *The age of alternative truths had officially arrived.*

The speed, scale, and virality of the lie is what made Sandy Hook a turning point. A few years earlier, a conspiracy so grotesque would have stayed locked in fringe forums alongside moon landing conspiracies and certainty that Elvis was alive and well and living in Tucson. But social media platforms had by then learned that outrage was good for business. YouTube's recommendation engine pushed the denial videos, Facebook groups multiplied like locusts, and hashtags spread the narrative to millions.

The Sandy Hook lie became the prototype of an emotionally explosive event repackaged into a shareable, monetizable, endlessly branching conspiracy ecosystem. And once that template existed, it was inevitable that antisemitic narratives, always lurking, always adaptive, would discover the format.

The strategy was now proven, when reality is unbearable, influencers can offer a more comforting fiction complete with villains. For many audiences, those villains would once again be Jews.

ISIS Cyber Propaganda

Between 2013 and 2016, ISIS unintentionally pioneered the playbook for modern digital radicalization. Its propaganda was groundbreaking in its understanding of how the internet could bypass borders, governments, and traditional media.

ISIS produced high-definition recruitment videos, cinematic battle montages, inspirational testimonials, and stylized graphic design that looked as if it belonged on a Hollywood storyboard. It understood virality much sooner than Western governments did. It used hashtags strategically, coordinated global Twitter storms, and built online communities that blended grievance, belonging, and violent ideology into a single potent digital cocktail.

Extremist groups everywhere, from white supremacists in Montana to Islamic terrorists in Gaza studied this success. ISIS proved that if your content was emotionally provocative, visually compelling, and algorithmically optimized, the internet would do the rest.

QAnon

The next wave of conspiracy culture, Pizzagate and its sprawling successor QAnon, absorbed many lessons from ISIS online tactics. Pizzagate began as a bizarre rumor about a Washington pizzeria and ballooned into a mythology of hidden cabals, ritualized child kidnapping, coded communications, and shadowy elites controlling the world. QAnon took this further, stitching together antisemitic motifs, including blood libel, secret societies, and global puppet masters, and draping them in the aesthetics of glossy infographics, YouTube explainers, and yoga influencer Instagram posts.

It was a bizarre fusion of medieval fearmongering and millennial lifestyle branding. The movement's so-called researchers, often suburban parents and wellness enthusiasts, cast themselves as truth-seeking warriors in a cosmic battle between good and evil. And while QAnon didn't always name Jews explicitly, the architecture of its narrative—secret global elites stealing children—is nearly indistinguishable from centuries old antisemitic mythology.

Jewish Journalists Targeted

One of the clearest signs that conspiracy culture had become intertwined with antisemitism was the way Jewish journalists found themselves at the center of coordinated harassment campaigns. When Julia Ioffe published a profile of Melania Trump in GQ magazine in April 2016, the backlash from the far right was swift, vicious, and unmistakably antisemitic. Her inbox filled with death threats, doctored images of her in a concentration camp gas chamber, and calls for her execution. It had nothing to do with the substance of her reporting. What mattered was that she was a Jewish woman visible in a national publication, and in the conspiratorial worldview taking shape online Jewish visibility is automatically recast as Jewish power, which is automatically recast as a threat. The attacks on Ioffe weren't random outbursts, but coordinated, meme-fueled swarm behavior made possible by anonymous platforms that turned hate into a participatory sport.

Other Jewish writers experienced different versions of the same phenomenon. Jeffrey Goldberg, editor-in-chief of The Atlantic, became the embodiment of the right wing fantasy of a media cabal, a coded phrase that dates back to centuries of antisemitic claims about Jewish control of the press. Even more telling were the attacks on Jewish contributors to Teen

Vogue, a magazine better known for skincare tips and prom dress guides than geopolitical controversy. Yet when Jewish women at Teen Vogue wrote pieces on politics, foreign policy, or social justice they were met with waves of coordinated antisemitic abuse. It was a reminder that in online conspiracy culture the specifics don't matter. To trigger a mob response, Jewish journalists didn't need to be covering Israel, finance, or national security; being Jewish and visible was enough. The only trigger was being Jewish and visible. The internet had created an atmosphere where Jewish identity alone was sufficient to activate the algorithms of hate.

How Social Media Supercharged Antisemitism

Social media platforms never set out to supercharge bigotry. They simply wanted to keep people scrolling. But in their quest to optimize engagement, they inadvertently discovered the universal psychological truth *that nothing holds attention like outrage*. Platforms tested thousands of variables and learned that posts provoking anger, fear, suspicion, or indignation performed better than all others.

Furious people stayed online and saw more advertisements, enriching the platform. The result was a perverse incentive system where the most emotionally explosive content, whether true, false, deranged, or hateful, was algorithmically elevated. Antisemitic conspiracies, always rich in emotional provocation, slotted perfectly into this economy. Hatred was profitable.

As the platforms were adapted to promote outrage, the line between fringe and mainstream evaporated. Ideas that once survived only in extremist forums and dark corners of the internet suddenly found themselves trending in Facebook groups for suburban parents, appearing in YouTube recommendation sidebars, or embedded in Twitter threads beneath cute dog videos.

Instagram's Explore page, designed to show you more of what you like, began to bring to the surface aesthetic conspiracy memes adorned with watercolor fonts and pastel backgrounds. The pipeline became painfully clear, someone would post a conspiracy laced infographic to a wellness group, where it would be reshared by perhaps a Pilates instructor, then reposted by her cousin, until ordinary people who had never heard of white nationalism were suddenly warning their followers about global elites pulling the strings. The platforms removed the walls that once kept these hateful messages contained.

At the same time, the internet's shift toward visual culture made propaganda easier than ever to consume. Gone were the days when antisemites were forced to demonstrate their commitment to the cause by reading a 40-page pamphlet about the Jew and the international order. Now, neo--Nazism was easy, even entertaining, to consume through a bite sized meme, an anime edit, a TikTok rant, or a polished YouTube video essay. Old lies were repackaged in vibrant colors, slick typography, and emotionally gripping narratives. Radicalization lost its ritualized form. Gone were the secret meetings, leaflet exchanges, and recruitment cells. Now there was just an algorithm suggesting one more video at 2:00 a.m.

All a user needed was a browser, a moment of loneliness, and a platform eager to feed whatever was ranking highest at a particular moment. Extremism had been fully decentralized. It no longer required leaders, organizations, or even a plan. It needed only the architecture of the internet, Wi-Fi, and an endlessly scrolling feed designed to keep people angry, awake, and online.

More Important Stories: Antisemitic Milestones of the Decade

One of the most chilling examples of how quickly antisemitic ideas can mutate in the digital age was the so-called "Parentheses Movement" of 2016. It began on the neo Nazi website *The Daily Stormer*, where users introduced a new stylistic code of wrapping a Jewish name in triple parentheses, like (((Goldstein))), to signal to other extremists that the person was Jewish and therefore deserving of harassment.

This signal was meant to function like a digital yellow star, a tagging system for intimidation in the algorithmic age. The meme spread across Twitter with astonishing speed, boosted by far right influencers and anonymous troll armies. Jewish journalists woke up to find their names suddenly encased in parentheses, triggering waves of death threats, doxxing attempts, and Photoshop imagery that invoked the Holocaust. The Anti-Defamation League quickly classified the triple parentheses as a hate symbol. The parentheses movement was a stark reminder that in the 21st century a punctuation mark can become a weapon.

The following year, the Unite the Right rally in Charlottesville showed how online hatred could erupt into real world violence. The rally was organized largely on Reddit and Discord, where thousands of users coordinated travel plans, created propaganda, shared maps of the march route, and designed the now infamous torchlit procession. Many of the participants only knew each other from online meme groups and private Discord channels where extremist ideology was packaged as edgy humor or historical truth telling. When they chanted *"Jews will not replace us,"* they weren't inventing a phrase, they were echoing a memeified version of the Great Replacement conspiracy theory, an allegation so old it dates back to the Egyptian Pharaohs (Exodus 1:9). The torches, helmets, and shields were analog, but the infrastructure that brought these people together, gave them confidence, and fed them ideology was deeply digital. Charlottesville was the moment the world realized that internet radicalization was no longer confined to screens.

Then came the Tree of Life massacre in 2018, the deadliest antisemitic attack in American history. The shooter was radicalized almost entirely online through platforms like Gab and 8chan, spaces intentionally built to evade content moderation and allow extremist content to flourish. His worldview was cobbled together from an online collage of white supremacist memes, slick conspiracy videos, lurid infographics, and endless forum threads insisting that Jews controlled immigration, manipulated elections, and threatened the white race. He frequently interacted with other extremists who praised violence, encouraged racist ideology, and treated antisemitic propaganda as entertainment. He didn't need to meet his ideological peers in a hidden barn, he met them in pixels, hashtags, and screencaps. When he entered the synagogue that morning, he carried not only a weapon, but an entire digital narrative that had been assembled for him by extremist communities.

These incidents revealed the terrible efficiency of online hate. What began as punctuation marks, memes, and message boards metastasized into street marches and gunfire. The Parentheses Movement showed how quickly digital signals can mobilize harassment. Charlottesville demonstrated how online networks can coordinate real world extremist action. The Tree of Life shooting proved the lethal consequences of algorithmically fueled radicalization. Together, they formed a tragic trilogy of the 2010s, a warning about how antisemitism evolves in the digital era, how fringe ideas move into mainstream consciousness, and how conspiracy culture, once unleashed, doesn't stay on the screen.

Modern Parallels: Why This Matters for the Post-Oct 7 World

When the world convulsed after the October 7, 2023, terror attack by Hamas on Israel, the digital chaos that followed didn't emerge from nowhere. The misinformation avalanches, the meme wars, the propaganda clips, the conspiracy floods, the university encampments, and the street protests were all constructed on a decade's worth of technological scaffolding. Everything the internet learned in the 2010s about making content sticky, emotional, and infinitely shareable snapped into place with brutal efficiency. The machinery was already built. The platforms had spent years teaching users to react instantly, trust visuals over facts, and treat information as entertainment.

Algorithms that had been fine-tuned for a decade to maximize engagement treated emotionally provocative lies as VIP content, pushing them to millions before fact checkers could even refresh their screens. Old hatreds now had the production value of a marketing campaign and the reach of a multinational corporation.

What made the post-October 7 environment uniquely destabilizing was the speed at which misinformation mutated. A single false video could be stitched, remixed, subtitled, and reframed hundreds of times in an hour. Narratives crossed platforms in seconds as a rumor started on Telegram leaped to TikTok, then to X, then to Instagram, and then to private WhatsApp groups that treated it as gospel. The volume was staggering, the emotional intensity was overwhelming, and the platforms were unprepared, though they should not have been entirely surprised. They had seen disinformation storms before, but never one that targeted Jews with such velocity and global uniformity. Now they saw the culmination of a decade of platform incentives that rewarded outrage faster than truth.

In the middle of this digital maelstrom were Jews, building vibrant online communities, sharing grief, organizing aid, telling their stories, and trying to correct falsehoods that swamped them. But they were also dodging weaponized conspiracies capable of reaching billions in seconds. Jewish voices, no matter how thoughtful or well-informed, found themselves outgunned by an attention economy that valued virality over accuracy, spectacle over context, and shock over humanity. The architecture of the 2010s had set the stage, and after October 7 the performance began. And the stakes were suddenly life and death.

A Jewish Conclusion to a Very Conspiratorial Problem

The internet of the 2010s enabled old antisemitic conspiracies to go viral. What Web 1.0 made possible, Web 2.0 made profitable. And the platforms that connected the Jewish world also connected the people who wanted to tear it apart.

In this new universe of algorithmic outrage and crowdsourced paranoia, Jews continued to do what they've always done—adapt, innovate, debate, organize, and, occasionally, retreat under the digital table until the shouting died down. But the shouting never really died down, it just evolved. The next wave, the era of TikTok, livestreams, and AI generation, would take everything to a new level of speed, scale, and surrealism.

Part 3

Social Media and the Algorithmic Age

Chapter 8

Algorithmic Psychology: Why the Internet Thinks Like a Brain on Fire

Understanding the modern digital world in all its speed, chaos, and emotional volatility does not require a degree in machine learning. It only requires a class in human psychology, preferably taught by someone who has spent time moderating a Facebook moms' group, a Jewish political forum, or a Reddit thread about Jewish culture.

The great secret of the internet is that, for all its computational sophistication, the platforms shaping our lives are built on a shockingly simple premise: *give humans what their brains already crave, just faster, louder, and without the inconvenience of reflection.* Algorithms don't create new desires, they intensify old ones for profit and market share.

That is what happens when the world's most powerful technology companies train machines not on truth or wisdom or dignity, but on the raw, unfiltered impulses of the human nervous system. That is also how those same impulses—fear, envy, outrage, tribal loyalty—map perfectly onto the psychology behind modern antisemitism.

The Human Brain, Reverse Engineered by Silicon Valley

Let's start with the basic fact that social media algorithms are not philosophical systems. They are attention-seekers. They don't care what is true, they care what keeps people staring. The goal is not enlightenment, but engagement. And the quickest route to engagement is to stimulate the amygdala, the brain's alarm system, with anything that feels threatening, infuriating, unfair, or humiliating.

In other words, *the algorithms are trained to seek emotional adrenaline.* Not information or nuance, or empathy—only adrenaline.

This is why a user feed rarely shows calm, thoughtful analysis. It shows someone yelling, a scandal, an accusation, a clip created to stimulate fury, or a meme that produces a sense of moral superiority in a single glance.

If Sigmund Freud and B.F. Skinner sat down together to design a platform that maximized human neurosis, insecurity, projection, tribal panic, and aggressive self-righteousness, they would have invented TikTok. And if Carl Jung wanted to summon the collective unconscious onto a global stage, complete with archetypes of villains, saviors, and shadow projections, he would have invented Twitter.

What Algorithms Want (and Why It's Not Good for the Jews)

Algorithms want only one thing – people's time. And the easiest way to extract time out of people's lives is to activate psychological instincts that kept humans alive in the Paleolithic world.

In this environment of emotional manipulation, antisemitism is premium fuel. Why? Because antisemitism has always been shaped by the same psychological drivers the algorithms intensify—fear of the hidden other, projection of societal anxiety onto a symbolic target, simplistic explanations for complex problems, a narrative with heroes and villains, and group bonding through shared resentment

Antisemitism is algorithm-friendly because it is simple, emotional, and endlessly adaptable. It's the ideological equivalent of a piece of clothing that somehow fits everyone.

The Algorithm Loves Outrage. Antisemitism Loves Outrage. A Match Made in Hell

Outrage spreads six times faster than joy, as neuroscientists have confirmed. And nothing triggers outrage like a story about a powerful conspiracy manipulating the world from the shadows. Enter the Jews.

Suddenly, medieval narratives find renewed traction and modern velocity. Old blood libels morph into supposed genocide infographics. Classic allegations of Jewish control become globalist puppet-master memes. The algorithm doesn't know these are centuries-old lies. It sees only the numbers of clicks, shares, and comments.

And so, the digital world becomes a vast amplification chamber whose emotional physics naturally elevate antisemitism, not because the platforms wanted hatred, but because hatred performs better.

Why Facts Lose: The Algorithmic Advantage of Emotion

The human brain is not wired to optimized facts, but to derive meaning. The easiest meaning to generate is threat. Algorithms exploit this fact by giving priority to content that spikes cortisol and elevates narratives that seemingly provide instant moral clarity (rewarding "us versus them" thinking) and suppress nuance, since gradation slows engagement. That is why well researched Jewish historical context loses to a 12-second TikTok clip claiming Israel invented colonialism. And why survivor testimony loses to an aesthetic infographic denying October 7 ever happened.

Algorithms are not referees, they are accelerants. They take whatever emotion exists and magnify it, then magnify the hashtag that magnifies the narrative that magnifies the grievance. When you magnify grievances long enough, it becomes identity.

The Three Laws of Algorithmic Psychology (That Nobody Wants to Admit)

If we were to rewrite Asimov's Three Laws of Robotics for social platforms, they might look like this:

- Law 1: The algorithm must optimize engagement, regardless of emotional harm. If it causes the user to cry, rage, obsess, or spiral, the feed knows it. And it gives more.

- Law 2: The algorithm must always amplify content that reinforces existing beliefs. Confirmation bias is not a bug; it is the feature that keeps people scrolling.

- Law 3: The algorithm must drive users toward more extreme content over time. Mild interest is fragile, intense emotion is sticky. So, the algorithm intensifies.

The internet is not a library. It is a recommendation engine trained on the physiology of stress.

The Vulnerability of Jews in an Algorithmic System

For Jews, algorithmic psychology has two especially dangerous effects—it transforms Jewish visibility into Jewish threat, and it converts Jewish trauma into viral content.

The more Jews speak, post, explain, or advocate, the more the algorithm exposes them to hostile audiences, tags them as political, builds an oppositional counter-audience, and funnels their content to people predisposed to hate them. This is why Jewish creators get mobbed, Jewish teens get flooded with threats, Jewish journalists get doxxed, and misinformation about Israel becomes the default explanation in millions of feeds.

October 7 was the most algorithmically accelerated trauma Jews have ever experienced. The platforms turned terrorism into a content category. The digital world learned how to monetize Jewish pain. That is algorithmic psychology at its most brutal, *human suffering as data*.

Why This Matters: The Architecture of Perception

The question as to why empathy for Jews evaporated so quickly after October 7 is not wrapped in mystery. It's algorithmic. The social platforms flattened context, gave priority to counter-narratives, and rewarded hostile aestheticized propaganda. The machines were promoting engagement, not morality, not truth, and most certainly not human feeling. Just engagement.

The Algorithms Are Not Evil, Just Indifferent. And Indifference Is Dangerous

Algorithms are not villains; they are amoral. They do not hate Jews. They also do not respect Jews, understand Jews, or care about Jews. They care about retention, clicks, dopamine, outrage, and scroll depth. Antisemitic instincts happen to dovetail with users' psychology which the system is designed to exploit.. Antisemitism is algorithmically compatible.

Pogrom Physics: How Jewish History Explains the Internet

Every communication revolution in Western history has struck the Jews like a tuning fork. Long before the first tweet, the Jewish people learned, through exile, persecution, and sheer adaptive genius, that new technologies do not merely transmit information: *they reorganize power.* They amplify the power of the masses, weaken old authorities, and create chaotic openings where hatred can scale up. Jews have lived inside the blast radius of these shifts for two millennia. To understand the modern internet, in all its beauty and brutality, there is no need to look at Silicon Valley. One can simply look at Jewish history.

The laws governing digital virality, speed, scale, replication, and emotional intensification, are the same social laws that governed pre-modern rumors and medieval mob psychology. What has changed is not the message but the medium. When a monk whispered a blood libel in the 12th century, it spread at the speed of horse or sermon. Today it spreads at the speed of a neural network optimizing for engagement. The difference is measured in milliseconds, but the underlying mechanism is ancient; outrage travels farther than nuance, fear outcompetes reason, and hatred finds its most receptive hosts among the bored, the angry, and the unanchored.

Jewish history is the empirical archive of how misinformation becomes violence. Consider the pogrom, a wave function of collective fear collapsing into physical destruction. Pogroms did not emerge from ideology alone; they followed predictable social dynamics. First, a rumor or accusation enters the system. Second, an authority figure fails to dampen it, or worse, amplifies it. Third, the crowd is saturated with it, reaching a critical mass. Fourth, the system tips from rumor to riot. The internet reproduces this model perfectly. It is a global machine for blurring the boundary between symbolic hostility and real world harm.

What Web 2.0 added was the long tail, millions of micro communities made viable by the death of geography. This seemed liberating, and in many ways it was. But it also recreated the medieval village, the scale at which conspiracy, gossip, and xenophobia have always thrived. A shtetl of strangers is still a shtetl. And when platforms are designed to boost the most emotionally charged content, the long tail becomes a long fuse. Anyone with a grievance can find a crowd, anyone with a prejudice can build a following, and anyone with a fantasy can manufacture a reality.

To study the internet without Jewish history is like studying nuclear reactors without studying Chernobyl. You learn the mechanics but not the risk. Jewish history is not only the story of oppression, but also a user manual for recognizing when systems built for connection become systems that reward chaos. Pogrom physics teaches that hatred is not a glitch, but an emotional energy that flows wherever the structure allows. Social media did not invent this energy, but removed the friction which deceleration requires. It built a civilization-wide accelerant.

The lesson is not despair but design. The internet does not doom us to perpetual tribal warfare any more than medieval Europe doomed Jews to perpetual persecution. Systems can be redesigned. Guardrails can be built. Incentives can be changed. Television, for example, was rarely exploited as a medium for the spread of antisemitism. That fact shows that regulation does not kill a medium but stabilizes it. Pogrom physics shows that hatred can be predicted, constrained, and outmaneuvered when society understands its patterns instead of denying them.

If the internet is the largest communication system ever built, then Jewish history may be the richest dataset for understanding its weaknesses. It teaches us that humanity is not inherently violent, but that humans under the wrong informational conditions can become violent very quickly. And it teaches that minority survival requires not only courage and resilience but an accurate model of how information spreads, mutates, and combusts.

The internet is not broken because people are bad. It is broken because the architecture rewards our worst impulses. Pogrom physics explains why. Jewish history explains how. Together they suggest a path forward, a digital world where connection does not require chaos, where speech does not require suffering, and where the lessons written in Jewish blood can finally be transformed into protections for everyone.

The Jewish Task in an Algorithmic World

While the 20th century taught Jews to understand geopolitics, the 21st century demands that Jews understand algorithms. What threatens Jews today is not only terrorists, mobs, dictatorships, conspiracists, or populists. It is the emotional physics of the internet itself.

The algorithmic mind is fast, reactive, tribal, moralizing, suspicious of complexity, allergic to context, and always hungry for conflict. That is to say, it thinks exactly like the parts of the human mind that antisemitism has always manipulated. Unless we understand that architecture and learn algorithmic psychology as fluently as our ancestors learned survival through adaptation, we will be fighting a 2,000-year old hatred with a 50-year old strategy in a world that changes every six months.

The digital universe is not neutral. It is psychological architecture which has already shaped the century. The question now is whether Jews can navigate it with the same ingenuity that carried them through every previous era of disruption. That's the real challenge, because the algorithms aren't slowing down.

Chapter 9

Meta: The Friendly Neighborhood Algorithm (with a dark side)

There was a brief, magical moment—let's call it 2010 to 2014 – when Facebook felt like the greatest invention in human history. People were able to see their college roommate's engagement photos, a coworker's dog, an aunt's favorite recipe, and a few mild political opinions that were generally scrolled past without any loss of faith in humanity. It was the digital equivalent of a bustling but friendly neighborhood with familiar faces, a little gossip, and nothing too alarming.

Then the algorithm grew teeth.

Between 2006 and 2018, Facebook didn't just become the world's dominant communication platform; it rewired how billions of humans understood information, community, identity, politics, and truth. It invented the News Feed, the Like button, algorithmic ranking, and social advertising at scale.

For Jews, this platform was both a digital promised land and an unmoderated minefield, often experienced within the same afternoon.

The Facebook Jewish Renaissance

Jewish communities embraced Facebook with enormous enthusiasm. Suddenly, every city had a Jewish Moms' group where people could discuss day school tuition, trade tips on where to find kosher mozzarella, and inquire about the halachic status of oat milk. Local buy/sell groups blossomed into informal Jewish marketplaces. Synagogues used Facebook for announcements and rabbis used it for Torah discussions. Activists found it useful for fundraising. The platform became the new Jewish community center, complete with gossip, debates, unsolicited advice, and the occasional heartfelt prayer request. American Jewish educators, who once relied on newsletters and email lists, suddenly had global visibility. A clever d'var Torah or spicy Talmud joke could reach thousands. A rabbi with a webcam could build a virtual congregation.

In Israel, Facebook was adopted with the same energy Israelis bring to their cafes, protests, and family dynamics. Political discourse, already famously intense, became hyper online. Israeli leftists, rightists, centrists, and the apathetic simply looking for a good argument, duked it out. Members of Knesset posted status updates that read like campaign speeches. Activists used the platform to organize rallies, sometimes within hours. Young soldiers posted satirical memes about army life (until the IDF leadership encouraged them to stop). From Tel Aviv to the smallest town deep in the Negev, community groups spun up for postings about lost pets, local corruption, traffic complaints, and the endless search for a reliable plumber. Secular Jews in Tel Aviv and yeshiva students in Jerusalem, who would never meet one another in the real world joined the same digital rooms where common interests prevailed. For a brief moment, it seemed that Facebook was becoming the virtual town square where every sector of society could gather to participate in the messy vibrancy of a nation in constant motion.

In many ways, Facebook strengthened Jewish life. It expanded access to resources, allowed mainstream and niche communities to flourish, and gave isolated Jews a sense of belonging. But the same infrastructure that made these connections possible also made antisemitic content frictionless, viral, and highly profitable.

How Facebook Became an Antisemitism Machine (Accidentally, But Effectively)

Few types of content generate outrage more efficiently than antisemitic tropes, Israel related misinformation, and conspiracy theories that happen to star Jewish protagonists.

Holocaust denial was allowed on the platform until 2020, protected under a misguided understanding of free speech. During those years, entire networks of denial pages flourished, reaching millions before Meta even noticed. Anti-Israel groups proliferated at astonishing speed. Moderation was inconsistent and under-resourced. Meanwhile, memes about Jewish cabals began circulating in political groups across the spectrum, ironically with many featuring Mark Zuckererg himself. On the right, they referred to globalist elites. On the left, they invoked corporate media manipulation or Zionist imperialism. In both cases, the architecture was unmistakably familiar.

The echo chambers were an additional accelerant. If a user clicked on one conspiracy theory, Facebook quickly offered up more of the same. The algorithm interpreted curiosity as endorsement. A user could start by watching a clip about global inequality and end up, after six scrolls down, reading about a Zionists plotting to conquer the world. This was the new ecosystem, antisemitism not as ideology, but as a content category. And content categories don't disappear, they scale up.

From Facebook Post to Real World Antisemitism

For years, people comforted themselves with the idea that online hate was just part of how the internet was, loud, obnoxious, unpleasant, but ultimately contained within screens. Facebook shattered that illusion. By the mid 2010s, it had become clear that what began as a meme, a comment, or a conspiratorial infographic on Facebook could easily spill onto streets, into synagogues, and across school campuses. The platform was no longer a passive mirror of public sentiment. It was an accelerant, taking the sparks of antisemitism that always existed and fanning them into open flames. The line between digital rhetoric and physical consequences narrowed until it effectively disappeared.

One of the earliest warning signs came out of Europe. In 2018, in Chemnitz, Germany, a small anti-immigrant protest organized on Facebook morphed into a full-fledged far right march complete with antisemitic chants and resulting violence. Investigators later found that many participants had been radicalized in local Facebook groups where conspiracy posts about Jewish financiers supposedly orchestrating migration had gone viral.

In France, during the Yellow Vest protests of 2019, Facebook pages circulated misinformation claiming Jewish leaders were behind rising fuel taxes. Within days, these conspiracies migrated into the streets, culminating in the harassment of Jewish philosopher Alain Finkielkraut by protesters shouting antisemitic slurs, some using lines lifted directly from Facebook posts.

The United States saw its own wave of platform-to-reality spillover. On May 18, 2021, during a round of Israel-Hamas fighting in Gaza, pro-Palestinian supporters in Los Angeles used social media to organize a protest that escalated into a physical attack on diners at the Sushi Fumi kosher restaurant, where patrons were beaten and harassed. Two days later, on May 20, 2021, 29-year-old Joseph Borgen was assaulted by a gang of pro-Palestinian activists who spotted him on West 49th Street after concluding a pro-Gaza rally nearby. Borgan was on his way to a pro-Israel rally when attacked. Throughout the United States and Europe Facebook served as source of incitement that led to the harassment, isolation, and physical assault of Jews, particularly post October 7.

At the communal level, even the soft underbelly of Facebook, like Jewish cooking groups, synagogue pages, and Jewish buy/sell boards found itself infiltrated. In 2020, a coordinated campaign by a British neo-Nazi Facebook network targeted dozens of Jewish groups around the world with floods of antisemitic memes. These weren't random trolls, but participants in a structured offline movement using Facebook as a staging ground. As a result, several synagogues reported vandalism within days of another, with perpetrators using imagery pulled directly from the Facebook meme dumps.

Perhaps the clearest example of Facebook's dangerous permeability came in 2021, when an antisemitic infographic falsely alleging a global Jewish media conspiracy went viral. The post was shared more than 300,000 times before being removed. In the aftermath, Jewish community centers from New York to Los Angeles reported real world vandalism featuring the exact same diagram printed out and taped to doors and bulletin boards.

Online hate didn't remain online. It slid off the screen, out of the groups, beyond the memes and into rallies, threats, vandalism, and violence. The platform had become a direct pipeline from misinformation to intimidation, from meme to menace.

When Facebook Broke the World (and What That Meant for Jews)

The turning point came during the Cambridge Analytica scandal in 2016 when the political consulting firm misused millions of Facebook users' data by harvesting personal information without consent via a personality quiz app in order to build psychological profiles for targeted political advertising. The firm collapsed in 2018.

This incident made it became clear that Facebook wasn't just a platform for sharing baby photos. Rather, it was a behavioral manipulation machine capable of influencing elections. Every political actor on earth took note. Extremists retooled their operations to exploit loopholes in the rules of engagement, and comment sections turned into digital gladiator arenas.

That same year, Facebook became implicated in the Rohingya genocide in Myanmar. United Nations investigators concluded the platform had allowed anti-Muslim hate to spread uncontrollably, fueling real-world violence. This became a case study for how Facebook handled, or rather mishandled, hate speech targeting minorities. Jewish organizations studied it carefully, because the mechanisms were identical to what they were witnessing—false rumors, hate memes, conspiracy posts, algorithmic amplification, and a moderation system incapable of keeping up.

Instagram: Quiet Engine of Jewish Identity

While Facebook was busy turning itself into the world's loudest town square, its younger sibling Instagram quietly became a laboratory for Jewish identity in visual form. Instagram was where Mizrahi and Sephardi Jews started reframing their story. Suddenly, feeds were full of Iraqi Jewish recipes plated like fine dining, Moroccan Mimouna spreads staged like fashion shoots, and Tunisian henna ceremonies shot in slow motion. For the first time, younger Jews from these communities could present their traditions not as footnotes to mainstream Ashkenazi culture, but as center stage material, bathed in good lighting and filtered through Valencia (a once-popular photo-editing effect). The platforms turned family memory into aesthetic activism, a daily correction to a Jewish narrative that had long treated them as background characters.

Jewish wellness influencers also found their niche on Instagram, mixing Torah quotes with yoga poses, posting about Shabbat as a tech detox, and turning concepts like teshuvah or emunah into carousel slides to be swiped through between latte art and cooking videos. Jewish fashion networks flourished, powered by women sharing tips on finding skirts that actually had pockets. For many women, especially in smaller communities, these accounts offered something their offline worlds rarely did, examples of people living visibly Jewish and visibly stylish lives without apology. Jewish life, for once, was not being filtered through the lens of stereotype, danger or trauma, but through color palettes and outfit grids.

But even in those curated spaces, antisemitism seeped in through the cracks. A post about Sephardi food could attract comments about European colonizers, or a modest fashion reel might be flooded with accusations about religious oppression. Wellness accounts discovered that the algorithm that loved their reflections on gratitude, also loved to bring to the surface conspiracy-tinted content, leading to Jewish creators frequently sharing screen space with posts about globalist elites and hidden cabals. Aesthetic platforms turned out not to be safe havens so much as quiet front lines, where Jewish visibility could be embraced in one second and attacked in the next.

The Dark Side Parallel: Meta as the Radio of the 1930s

Just as radio in the 1930s created mass scale messaging that could be hijacked by fascists, Meta's Facebook and Instagram created a digital public square capable of being hijacked by extremists. The difference is reach and velocity. A hateful radio broadcast reached thousands, a viral Facebook post reaches millions, within hours. And when the algorithm decides that a piece of antisemitic content is engaging, it sends it around the world like a chain letter written by a racist bot.

Meta showed the world that technology doesn't need to intend harm to Jews in order to bring it about. All it needs is a design that rewards emotional extremism, a moderation system that can be easily overwhelmed, and billions of users who don't know fact from fiction (and don't care).

For better and for worse, Facebook and Instagram did what every major communication revolution does, they connected Jews, and they connected the people who hate Jews. The difference this time was scale. And the next platforms would scale up everything even more.

Chapter 10

YouTube: the Unseen Giant

If Facebook was the neighborhood bulletin board, YouTube was the giant glowing TV quietly bolted to the wall of civilization. It never boasted of being a character in the information ecosystem, but rather just sat there, endlessly auto playing, while three billion people a month wandered in and out of its doors.

For Jews, YouTube became many things at once; a global classroom and a digital beit midrash, but also an antisemitic funhouse, and a conspiracy theme park with no closing time. It was where kids learned Alef-Bet songs, teens got called slurs in gaming chats, and, despite heroic efforts to lift the platform toward something resembling truth, adults nonetheless fell down the Middle East rabbit hole.

And for a long time, almost nobody realized that this quiet video site, more than any other single platform, was training the world how to think.

The Video Library That Ate the World

YouTube was launched in 2005 as a place to upload home videos, bootleg music clips, and footage of people falling off skateboards. By 2010, it had become the world's largest educational archive, music catalog, and propaganda distribution system all at once.

YouTube had two defining features that made it uniquely powerful. First, the Recommendation Engine, that famous sidebar and autoplay queue that pointed out what users might like based on what they were viewing, and then gently escorted them from a harmless clip into progressively more engaging content. "Engaging", of course, is Silicon Valley-speak meaning emotionally charged enough to keep viewers watching, even if they are supposed to be doing some other priority tasks. Second, the low barrier to entry.

Anyone, anywhere, with a phone or a webcam could become a broadcaster. You didn't need a TV network, a publisher, or a studio. You just needed the ability to talk at length into the internet void and occasionally remember to ask viewers to click "Subscribe" and "Like". For Jewish life, this was transformative. For antisemitism, it was a dream.

The YouTube Beit Midrash: Rabbis, Rebbes & Ring Lights

Jews are not known for being shy about talking. Give a rabbi a camera and Wi-Fi, and there will be a channel. By the late 2000s, almost every variety of Jewish thought could be found on YouTube. There were Hasidic storytellers sharing Breslov teachings in English, modern orthodox rabbis teaching daf yomi with annotated slides, and reform rabbis answering questions from cultural Jews about cheeseburgers. There were Chabad short clips about tefillin, Shabbat, and the mystical meaning of coffee, academic scholars turning their lecture notes into mini courses on biblical criticism, and Israeli teachers giving Hebrew lessons with cartoon graphics and very enthusiastic hand gestures.

This was the first platform where Jewish education scaled up globally without the mediation of institutions. For a people whose identity is wired to study, this was intoxicating. The People of the Book could now also be the People of the Playlist. And for a brief, hopeful moment, it felt like YouTube might be the great equalizer, the place where Jewish wisdom could flow as freely as pet videos.

And Then Came the Gamers

Unfortunately, YouTube was not just a beit midrash. It was also a massive gaming stadium where teenagers screamed into microphones while playing *Call of Duty*, discovering along the way that antisemitic slurs sounded extra edgy when shouted in surround sound.

YouTube gaming channels became a cultural universe unto themselves. Streamers were idols. Their jokes, slang, and insults were copied by millions. If a popular gamer casually dropped the word *Jew* as an insult, the phrase embedded itself in the language of their fans.

For a whole generation, antisemitic language slipped in as just another meme, another edgy bit of online vocabulary. It wasn't old men in trench coats handing out fliers anymore. It was a 19-year-old with a headset and a sponsorship deal, laughing while he called someone a Kike to impress the group.

YouTube became the place where antisemitism wasn't always ideological, it was also ambient. Less like a manifesto and more like background music.

The Conspiracy Rabbit Hole, Brought to You by Autoplay

To understand YouTube's role in 21st century antisemitism, we have to understand the rabbit hole. Here's how it has worked for millions of users; a user begins, for example, with a perfectly reasonable video, say one called, "*The History of the Israeli–Palestinian Conflict.*" The algorithm, noticing the interest in geopolitics, might then recommend another video, this one called, *"The Untold Truth About the Middle East,"* a little spicier, a little more dramatic. If the user sticks around a bit longer, suddenly they're being urged to watch a video titled, *"Zionism and Genocide."* At no point did the user request a video about Zionism. YouTube simply walked the user from mild curiosity to classic antisemitic conspiracy theory because each step along the path looked potentially engaging.

The same thing happened with 9/11 content. A user might click on a documentary about the attacks. The sidebar would then suggest videos asking so called "hard questions" about Building 7. A few clicks later, that user is watching a video connecting 9/11 to the Mossad, Zionism, and international Jewry, all presented in slick graphics with ominous music. Old blood libels with a new soundtrack.

Holocaust denial also found fertile ground. For years, searches for Holocaust on YouTube could produce denial videos, self-styled revisionist lectures, and staged debates questioning whether gas chambers existed—all framed as *just asking questions*, as Joe Rogan might say.

The platform eventually took steps to downrank or remove some of this content, especially after public pressure and academic studies showed just how often users were being recommended extremist videos. But the pattern was already established, YouTube's core design made it fantastically adept at turning curiosity into radicalization. And because antisemitic conspiracy theories are the Swiss army knife of paranoia, they fit perfectly into every rabbit hole on offer.

When Long Form Radicalization Met Real World Violence

If you look at the biographies of several high profile extremists in the last decade, including some who targeted Jews, you find they almost certainly have a catalogue of hours and hours of YouTube videos. They produced video essays about supposed globalism, immigration conspiracies, cultural Marxism, replacement theory, degeneracy, and the pending collapse of Western civilization. Their videos were often slick, articulate, and framed as if well--meaning skeptics are simply asking forbidden questions.

Jews, in this worldview, appeared as symbols more than people. They were the cultural engineers, the media gatekeepers, and the architects of globalism. YouTube wasn't the only platform feeding these ideas, but it was where the ideas were given production value.

A particularly insidious aspect of YouTube radicalization is that it feels like self-education. Unlike a live TV broadcast, where viewers passively receive information of the broadcaster's choosing, YouTube invites viewers to click, select, and assemble their own programming. This illusion of self--guided learning is powerful. If a user later encountered mainstream sources contradicting what they've seemingly learned from hours of YouTube videos, it was easy to dismiss those sources as towing the line of the international cabal.

For antisemitic narratives, this was a gift. It allowed age-old lies to masquerade as fresh discoveries. The act of exploration became self-validating and bestowed nearly unquestionable legitimacy to what was viewed. A generation became convinced they had uncovered truths about Jewish power that had been hidden by schools, journalists, and governments.

So Jewish YouTubers improvised. They learned thumbnail tricks, used better titles, adjusted pacing, and inserted hooks into the first ten seconds. Some reluctantly adopted YouTube's own tactics in hopes of bending them back toward something good. They became both teachers and reluctant marketers, trying to sneak vitamins into the algorithm's diet of sugar. It wasn't a fair fight, but then, Jewish history is mostly a long compilation of unfair fights.

A Jewish Conclusion to a Very Visual Problem

YouTube rarely gets the spotlight in conversations about antisemitism. People talk about Twitter/X because it's loud, about TikTok because it's dynamic, and about Facebook because their uncle is on it. YouTube is more like infrastructure, so omnipresent we barely notice it.

But in the story of Jews and technology, YouTube is the unseen giant. It quietly taught generations how to learn, how to argue, and how to see the Jews, sometimes as teachers, sometimes as punchlines, and often as villains in stories with a suspiciously familiar plot.

For Jews, YouTube brought the extraordinary gifts of unprecedented access to Torah, culture, language, and global connection. It also brought new forms of vulnerability, embedding antisemitic narratives into the architecture of online learning itself.

As always, the pattern repeats: *the same machine that preserves Jewish memory also spreads anti-Jewish myths, and the same tool that can teach us empathy can also lead us to hatred.*

And just as Jews were getting used to this long form, desktop anchored video world, a new platform quietly began rewriting the rules. It was built for phones not laptops and it traded half hour lectures for 12 second clips. It didn't bother pretending to be a library. It wanted your feelings, not your attention span. That platform, TikTok, was about to turn the algorithm into something more intimate and emotional, and, for Jews, far more volatile.

Chapter 11

TikTok: The Teen Algorithm That Accidentally Turned Political

If Facebook was the messy digital neighborhood and YouTube was the global rabbit hole, TikTok arrived like the algorithmic child prodigy who skipped grades, mastered propaganda distribution, and accidentally became secretary general of world opinion.

Launched globally in 2018, TikTok redefined virality. By 2020, teens, adults, governments, and intelligence agencies were all trying to figure out how an app designed for dance routines and lip syncing had become one of the most powerful political communication engines on earth.

TikTok's genius, and its danger, lies in its For You Page, a personalized conveyor belt of emotionally charged, visually rich content tailored to keep you scrolling with the manic focus of a man searching for his lost car keys. The algorithm's goal is to show whatever makes the user's brain light up, be it a tragedy, a joke, a cute dog wearing sunglasses, or an antisemitic blood libel, preferably in under 12 seconds. TikTok is, in effect, the world's most efficient empathy manipulation engine, optimized for speed not truth.

For Jewish creators, TikTok was initially a bustling hub of culture, humor, ritual, and identity. But alongside all this warmth came something colder, sharper, and far more predictable—antisemitism infused into trends, soundtracks, memes, duets, and stitched videos. As with some earlier platforms, TikTok became the first place where many young Jews encountered antisemitic hate for the first time. The platform's design made it easy, so anti-Jewish stereotypes, simplified or false narratives about Israel, Holocaust jokes, and blood libel fantasies spread like wildfire.

After October 7, 2023, the floodgates opened. Pro-Hamas content surged globally. Teen influencers posted so-called *"explainers"* of Middle East geopolitics with the confidence of tenured professors and the historical literacy of an eggplant. Emotional misinformation outpaced verified reporting by a factor of thousands.

Holocaust denial adapted to the format, too. There were the infamous Holocaust filters, videos asking viewers to imagine themselves as camp victims, and POV videos that trivialized genocide in the name of aesthetic storytelling. TikTok's format—fast, visual, decontextualized—turned trauma into trends and geopolitics into vibes.

In a sense, TikTok revived an old propaganda truth that speed beats nuance. The Nazis knew it in the 1930s, using fast, emotional visuals to bypass reason. TikTok replicated the technique not ideologically, but structurally. Compress anything into 12 seconds with a dramatic string track behind it, and a piece of micro propaganda will have been created, whether intentionally or not.

From TikTok to Real World Antisemitism

For years, people assumed TikTok was just for kids, a place for dances, trends, idiotic challenges, and makeup tutorials. After the October 7 Hamas massacre, suddenly the world realized that the platform shapes beliefs, identity, and politics on a planetary scale, especially among young people.

And as with every major communication revolution, online antisemitism migrated offline with alarming speed. In TikTok's world, antisemitism isn't a fringe problem, it's content. And content always wants to become action.

The "Jewish Privilege" Harassment Wave

From mid-2020 to 2021, TikTok users began circulating videos alleging Jewish privilege, which implied that Jews benefited from structures of racial oppression. The result was that Jewish high school, and college students were confronted on campuses with the phrase *"check your Jewish privilege,"* often by peers who had seen the trend on TikTok. Several Jewish teens reported being cornered in bathrooms or hallways by fellow students parroting lines from viral TikTok videos. In Los Angeles, a Jewish teen was shoved and taunted with *privileged Jew* slogans, language lifted verbatim from TikTok clips. These were the first major indicators that TikTok content wasn't just influencing conversation. It was shaping behavior.

Copycat Attacks from Viral Holocaust Challenges

In 2020 TikTok introduced filters and point of view (POV) storytelling formats, creating the immediacy of making viewers feel they actually stand in the shoes of the person posting the video. Within a short time, some users created grotesque videos portraying themselves as Holocaust victims. This fueled an outbreak of antisemitic bullying in schools. Multiple Jewish middle schoolers reported classmates mocking them by reenacting TikTok Holocaust videos in hallways. A Jewish student in the United Kingdom described peers giving Nazi salutes. Teachers across the U.S. reported needing to intervene in antisemitic role playing sparked entirely by TikTok trends. The digital trivialization of genocide became physical harassment.

The Viral Anti-Israel Sound That Led to Real World Threats

During the 2021 Israel-Hamas conflict, a TikTok audio, with dramatic music and maps filled with misinformation claiming Zionists control the world went viral. Offline consequences followed quickly. A synagogue in Toronto was vandalized, and investigators tied the graffiti to a TikTok video that had recently gone viral. In New York, Jewish teens reported strangers yelling at them lines from videos while they waited on a subway platform. A Jewish teacher in Florida received anonymous threats parroting the exact wording of a video. TikTok had turned extremist audio into an ambient cultural script.

Campus Activism Scripted by TikTok

After October 7, 2023, TikTok became the primary engine for student activism, and the primary engine for antisemitic misinformation. The platform popularized simplistic narratives portraying Jews or Israelis as absolute villains. At colleges across the U.S., these narratives jumped offline. Students in multiple universities reported being surrounded and shouted at with slogans that originated in TikTok explainers. Protest organizers at several campuses used TikTok videos as de facto training materials, leading to chants and behaviors rooted more in TikTok rhetoric than in any political ideology. TikTok, in addition to influencing opinion, was choreographing protest.

The "From the River to the Sea" Explosion

After the Hamas attack on October 7, the resurgence of the chant "from the river to the sea"—a call for the elimination of the State of Israel—was powered massively by TikTok's algorithm. The chant quickly jumped offline. In cities like London, Paris, Madrid, Belfast, Berlin, Sidney, and New York, marchers repeated the slogan after first encountering it through TikTok influencers. Jewish pedestrians reported being confronted with the phrase, sometimes aggressively, by strangers. A Jewish-owned cafe in Toronto reported vandalism with the phrase spray painted across its windows just days after a regional TikTok spike. A chant that many users didn't even understand, ignorant as they were as to which river and which sea, had gone from screen to sidewalk.

A Jewish Conclusion to a 12 Second Problem

TikTok didn't set out to become the world's most powerful distributor of political emotion or a conveyor belt for 12 second conspiracy theories. It simply discovered that the shortest path to engagement is through these types of posts. And once that door opened, antisemitism, already well practiced at disguising itself in whatever form a society finds most agreeable, slipped right in.

Just like every technological revolution that came before, from the printing press to Facebook, TikTok left Jews doing two things at once—-building vibrant new forms of digital Jewish life, and simultaneously ducking waves of algorithmically amplified hate. The duality was familiar, the speed was not.

TikTok was only the beginning of the short-form propaganda era. Twitter/X was even faster, louder, more combative, and somehow even less contextual. Twitter/X took everything TikTok normalized and injected it directly into the bloodstream of global discourse. If TikTok was where teens learned geopolitics through vibes and violin tracks, Twitter/X was where adults turned those vibes into public meltdowns, political movements, and real time rage cycles. It was the place where antisemitism didn't just spread, it trended.

Chapter 12

Twitter/X: Where Public Discourse Goes to Burn

If Facebook was your aunt's living room and TikTok was the world's teenage diary, Twitter, later rebranded X, was always the digital equivalent of standing in the middle of Times Square with a megaphone and hoping no one threw a bottle at you. From 2009 to 2022, Twitter served as the world's political newsroom, town square, comedy club, protest hub, breaking news wire service, and ongoing global nervous breakdown. It was where journalists lived, politicians fought, comedians tested material, and academics posted their neuroses for all to see. And then Elon Musk bought it.

With the flip of a switch (and $44 billion), the world's messiest platform became even more chaotic. Musk weakened moderation systems, reinstated previously banned extremists, turned verification into a purchasable vanity accessory, and personally began engaging with conspiratorial content. The platform had always been combustible, and now it had a leader flicking matches onto the gasoline.

Jewish Twitter: A Global Salon, a Support Network, and a Contact Sport

For Jews, Twitter was once again a contradictory paradise. It was one of the few places where someone could watch a Talmud scholar debate a comedian, a rabbi argue with a secular Israeli journalist, and a New York deli owner weigh in on geopolitics. Jewish Twitter became a cultural zone unto itself. Writers like Bari Weiss, Ben Shapiro, and Batya Ungar-Sargon shaped discourse. Israeli politicians, from Naftali Bennett to Yair Lapid, used the platform like a live press briefing. It was a global shtetl with Wi-Fi.

At the same time Twitter was also one of the most hostile environments for Jewish women. Harassment campaigns surged around journalists like Julia Ioffe, Sarah Jeong, and Caroline Orr. "*Go back to the gas chambers*" became a disturbingly common message sent by anonymous accounts tweeting with all the courage that only anonymity can provide. During conflicts involving Israel, Jewish voices were routinely hit with waves of coordinated abuse, sometimes from genuine users, sometimes from bots, and sometimes from state sponsored networks in Iran, Qatar, Turkey and Russia.

How Twitter/X Fueled Antisemitism

Twitter's design was uniquely suited to amplify hate. The trending topics list was essentially a stage, and whoever shouted loudest from the crowd got handed the microphone. Outrage was the platform's native language. Conspiracies spread quickly because they were short, emotional, and easy to retweet. Bots helped do the rest.

After Musk's takeover in 2022, hate speech increased sharply. Extremists previously banned for incitement, neo-Nazis, white supremacists, and QAnon influencers, were all reinstated. State sponsored actors also got creative. Iran and Qatari linked networks pushed anti-Israel and antisemitic propaganda. Russian troll farms seeded Jewish conspiracy narratives into Western political conversations. The platform often became indistinguishable from a Cold War operations center staffed by meme lords. Moreover, when Musk himself interacted directly with conspiratorial posts, including the infamous claim about Jewish funded NGOs controlling immigration, he didn't just amplify them, he legitimized them. Measuring extremism on Twitter became harder simply because the most followed account on the platform (Musk's) was sometimes the source of the problem.

Important Stories & Moments from the X Ecosystem

Kanye West's Antisemitic Spiral

In 2022, Kanye West, rapper, fashion mogul, former spouse of Kim Kardashian, and perennial generator of public spectacle, chose Twitter as the stage for his most destructive performance yet. In a late night burst of posting, he announced he was going "death con 3 on JEWISH PEOPLE," a phrase that combined blatant antisemitism with a typo so confusing that linguists and rabbis alike had questions.

Within minutes, the tweet exploded across the platform. This wasn't an obscure influencer ranting into the void, but one of the most recognizable celebrities in the world telling his 30 million followers that Jews were a threat. The spread was instantaneous and the reaction volcanic.

Almost immediately, Twitter filled with hashtags like #YeIsRight, which trended globally, sometimes for hours at a time. Millions watched as antisemitic accounts, newly emboldened, used Kanye's words as validation for their own conspiracies. Far right influencers declared the tweet *"the moment the truth comes out,"* Neo-Nazi accounts praised Kanye as a truth teller. Even people who had never engaged with antisemitism before found themselves swept into the discourse through quote tweets, memes, and algorithmically boosted content. The platform became an echo chamber of celebrity amplified hate, with Kanye at its center.

The impact didn't stay online. In Los Angeles, a group of neo-Nazis hung a banner over the 405 freeway reading "Kanye is right about the Jews." Jewish families reported strangers shouting the same phrase out of car windows. Synagogues in Miami and Austin were vandalized with slogans and symbols referring to Kanye's tweets. Jewish celebrities, many of whom had never commented publicly on antisemitism, began hiring additional security after receiving threats that used Kanye's language verbatim. TikTok and Instagram copied the rhetoric, but Twitter was the birthplace, the accelerant, and the proof of concept for how quickly antisemitism can ignite globally when a famous person lights the match. It was the perfect demonstration of how celebrity influence, platform algorithms, and preexisting antisemitic narratives can combine to form a cultural event.

#IsraelUnderFire vs #FreePalestine

The dueling hashtags, #IsraelUnderFire and #FreePalestine, turned Twitter/X into digital trenches every time violence erupted in the Middle East. These hashtags weren't just descriptors, but battle flags under which users, instead of simply posting content, enlisted in digital battle. Each tag became a curated echo chamber reinforcing a worldview and narrative arc, often emphasizing a chosen villain. Once a conflict broke out, the platform's algorithms shifted into high gear, pushing the most emotionally charged content to the surface. The result was a real-time propaganda storm, with people owning a smartphone thinking they were directing it

Misinformation circulated with astonishing speed. It didn't matter if it was a video from Syria or Gaza or a clip from a *Call of Duty* video game. Every recycled image came with captions written with absolute certainty, usually by people who could not have pointed to Israel or Gaza on a map. Users posted dramatic footage without context, insisting that the morality of their side justified the sloppiness of their sourcing. Since the algorithm rewards engagement, not accuracy, the most misleading content often reached the widest audience. On Twitter, conviction beat verification every time.

Meanwhile, journalists trying to provide verified information found themselves overwhelmed by the torrent. Jewish journalists in particular became lightning rods for rage, no matter what they reported or how carefully they sourced it. Merely having a Jewish name or a Hebrew speaking profile could trigger swarms of replies filled with slurs, conspiracy accusations, or Holocaust references. The conflict wasn't just geopolitical. On Twitter, it became personal, and was aimed at individual users.

Israeli users faced coordinated harassment. Every tweet from an Israeli account, no matter how mundane, be it a photo of a dog on a Tel Aviv beach or a video of a Negev sand dune, could attract replies insisting the account holders were complicit in genocide. Bots and state linked troll networks amplified hostile responses. Iranian, Qatari, and Russian influence operations pumped out hashtags, memes, and misleading graphics designed to inflame division. For many Israeli users, opening Twitter felt like walking into a wind tunnel of rage designed specifically to destabilize them emotionally. In some cases, Twitter became a substitute for the battlefield, a place to inflict psychological wounds while the physical war was happening miles away.

The dynamic extended far beyond Israelis and Palestinians to global Jewish users who became collateral damage. Jews everywhere, from London to New York, Brussels to Sydney reported a surge in online harassment during every conflict cycle, even if they had never tweeted about Israel in their lives. Jewish academics found lectures interrupted by online mobs. The conflict's hashtags were a funnel for the moment they trended and antisemitism spiked everywhere, even in cities where the only Middle Eastern conflict was between rival falafel shops.

In the end, Twitter/X became a pressure cooker, and Jewish users, both in Israel and in the diaspora, became the steam valve. Clearly, the platform wasn't designed for nuance, historical context, or moral complexity, but for acceleration. Each hashtag was a spark thrown into a dry haystack of identity politics, nationalism, trauma, and misinformation. The fires spread instantly, and each time Jews found themselves caught between worlds, targeted not only for their opinions but for their very existence. On Twitter, the conflict was constant, the outrage global, and the casualties emotional as much as informational.

A Jewish Conclusion to a Very Musk-Propelled Problem

In the end, Twitter/X didn't destroy public discourse, it merely revealed how combustible the discourse already was. It showed how fast a rumor can harden into belief, how easily harassment can masquerade as activism, how quickly millennia old hatreds can be repackaged into trending topics, and what happens when a social platform owner behaves irresponsibly.

For Jews, the platform was a paradox, a place to build community and fight misinformation in real time, but also a place where antisemitism could erupt with no warning, having a swarming effect. The sheer velocity of Twitter meant there was no time to breathe, no time to contextualize, no time to think, only time to react.

But Twitter was never just a platform. It was an arena, where identities were forged, challenged, attacked, and defended in public view. It transformed users into avatars, arguments into moral crusades, and disagreements into tribal warfare.

That raises the real question: if Twitter showed us how fragile discourse is, what happens in a digital world where identity itself becomes the battlefield, where influencers double as political actors, where activism merges with performance, and where tribal belonging is defined by hashtags, aesthetics, and algorithms?

Chapter 13

Identity Wars: Influencers, Activism, & Digital Tribes

By the mid-2010s, a new digital order had formed, governed by influencers, activists, and content creators, rather than politicians or editors. Identity became the currency of moral authority and suffering became a credential. Users were encouraged to figure out exactly which global tribe they belonged to, ideally using a series of color coded slides explaining the world in three bullet points.

This era didn't begin with conflict. It began with genuine liberation movements like Black Lives Matter, # MeToo, disability justice, indigenous activism, queer liberation, and a renaissance in anti-racist scholarship. These movements used social media to challenge silence, expose injustice, and mobilize millions. But as their strategies became mainstream, a new style of digital activism emerged built on viral storytelling, simplified victim/oppressor binaries, emotionally charged narratives, and shareable morality. In the algorithmic ecosystem, complexity died quickly. The more straightforward the identification of the villain, the more viral the post. The more dramatic the oppression narrative, the more likes. Intersectionality, the brilliantly nuanced framework coined by legal scholar Kimberle Crenshaw, was flattened into social media-ready shorthand for one's place in the moral hierarchy. This is where things got complicated for Jews.

Jews in the Algorithmic Sorting Hat

Jewish identity has always been resistant to neat categorization, so when the internet began dividing the world into identity groups, Jews became the glitch in the matrix. Jews were white, except when they absolutely weren't. They were a minority but not always perceived that way. Identity spaces online, though, had no patience for ambiguity. The algorithm needs clarity.

The fact that half the world's Jews are brown or olive-skinned descendants of communities expelled from Islamic countries was lost in translation. Online activism rarely had a category for Jews who fled Arab pogroms, or Persian Jews forced to escape on forged passports, or Soviet Jews imprisoned for studying Hebrew. The model allowed only two boxes, and the Jews didn't fully fit in either. And so Jewish identity got flattened into something inaccurate—Jews *are white and privileged and so therefore they are the oppressors.*

Even worse, Jewish trauma, so central to Jewish identity, was frequently erased in digital spaces. Posts claiming that Jews benefit from proximity to whiteness circulated widely in Diversity, Equity, and Inclusion training and in social media infographics that treated Jewish history as a footnote. The long arc of antisemitism—slavery, exile, ghettoization, expulsion, and genocide—was rewritten as a minor subplot in a broader story about white privilege. Jews found themselves caught between categories, holding their trays awkwardly in the cafeteria line of digital identity politics.

Influencers, Tribes & the Pressure to Declare Allegiance

Influencer activism not only classified people, but also asked them to pick teams, and to do so publicly and in the performative style of the algorithm. Suddenly Jewish content creators, cooks, comedians, and educators found themselves asked to explain their stance on Zionism, colonialism, anti-racism, and Middle Eastern geopolitics, often by followers who had never asked them a question about Judaism or Jewish life before. *Are you a Zionist?* became the litmus test question of the decade, demanded not by Middle Eastern scholars, but by strangers in the comments to a video rating the quality of a New York pizza.

Within these digital tribes, solidarity became highly conditional. Jewish creators who aligned perfectly with other progressive causes still found themselves shut out when they raised concerns about antisemitism. Accounts that typically posted about anti-hate initiatives struggled to include Jews in their frameworks. And when Jewish influencers posted about the rising antisemitic violence they faced offline, they were often met with silence, or worse accusations that they were distracting from more important issues.

The Algorithmic Rebirth of Old Narratives

The most dangerous outcome of digital identity wars was the way they revitalized old antisemitic narratives in the language of contemporary activism. Claims that Jews are privileged, cosmopolitan elites, or global power brokers gained traction in online spaces that had no interest in medieval history but every interest in moral simplicity.

The term Zionist became a catch-all insult for Jews broadly, even for those who had never set foot in Israel. In online debates, anti-Israel rhetoric often slid, with astonishing speed, into classic anti-Jewish generalizations. The line between political critique and ethnically targeted hatred blurred into irrelevance.

A Few Flashpoints That Defined the Identity Wars

During the 2020 racial justice protests, a debate broke out in activist spaces over whether antisemitism should be included in anti-hate frameworks. For many Jewish activists, this was a bewildering turn. They arrived at marches and virtual town halls prepared to explain, often patiently, that white supremacists do not view Jews as white at all, quite the opposite. Neo-Nazis, the Klu Klux Klan, and other extremist groups consistently classify Jews as impure, or non-white in their racist taxonomy. Yet in progressive spaces, Jewish participants were told that antisemitism wasn't central enough, urgent enough, or structurally relevant enough to be included.

This created a strange dynamic whereby a group historically targeted was being told it was too privileged to count as a target of bigotry under the new paradigm. Jewish activists found themselves doing a kind of historical gymnastics, citing everything in Jewish history—slavery, conquest (occupation), forced diaspora (dispossession), forced conversions (ethnic cleansing), medieval expulsions (displacement), ghettoization (apartheid), and the Holocaust (genocide)—in order to confront the surreal inversion of the Jewish historical experience. They even referred to current FBI data and hate-crime statistics showing alarming rates of anti-Jewish behavior, but in identity circles built around rigid hierarchy models, all those arguments were most often dismissed outright.

This disconnect wasn't rooted in malice but came from the simplified frameworks of digital activism that were built to explain systems of oppression, rather than the intricacies of Jewish history. Activist spaces had clear categories for understanding racism, colonialism, xenophobia, and Islamophobia. But antisemitism, an ancient, shapeshifting ideology that doesn't map neatly onto race, class, or national origin, didn't fit comfortably into these structures. It was the trick question in the algorithm, the exception that messed up the chart.

Meanwhile, antisemitic incidents surged nationally with synagogue shootings, harassment on campuses, vandalism, online conspiracies, and threats. Moreover, the gap between DEI ideology and Jewish lived experience grew even wider after October 7, 2023.

In many progressive spaces, the reluctance to view Jews as victims of oppression is deeply entangled with political attitudes toward Israel. For a generation shaped by social media activism, Israel is often framed in purely colonial terms as the powerful oppressor in a simplified global morality play. And because Israel is widely (and somewhat incorrectly) treated as synonymous with all Jews everywhere, that political framing bleeds directly into perceptions of Jewish identity. The logic becomes, if Israel is powerful, and Jews are associated with Israel, then Jews, by extension, must be powerful too. This collapses the vast diversity and vulnerability of Jewish communities into a single politicized stereotype that leaves no room for Jewish trauma, Jewish marginalization, or Jewish persecution in the diaspora.

As this logic hardens, it creates a kind of conceptual blind spot. In progressive frameworks shaped by victim-oppressor binaries, there is little space for a group that is both historically oppressed and associated with a country perceived as militarily strong. The long arc of Jewish suffering is overshadowed by a hyperfocus on Israel as the defining element of Jewish identity. This leads to the bizarre situation in which Jews are told that they can't experience racism because they are linked to Israel, which they view unfavorably. Political disagreement with Israel morphs into a broader denial of Jewish vulnerability and rights everywhere.

This dynamic is made even more volatile online, where people treat Israel as a monolithic villain and Jews as a monolithic extension of that villainy. In this environment, acknowledging Jewish victimhood can feel, within certain activist circles, like betraying solidarity with Palestinians. As a result, Jewish pain is framed as politically inconvenient, Jewish history is minimized, and Jewish fear is dismissed. What emerges is not principled critique of a government, but a worldview in which Jewish suffering is rendered illegible unless it can be weaponized against Israel itself. The result is the painful irony that the very communities committed to fighting oppression often fail to recognize one of history's most consistently oppressed peoples, simply because their distorted view of a nation-state thousands of miles away complicates their ideological map.

For Jewish users, the consequences were immediate. Online identity politics didn't just misinform or marginalize, they created moral and identity tribes. Once a tribe formed around a narrative, no matter how inaccurate, it acted as a fortress against dissent. Jewish voices offering nuance were labeled apologists or colonizers. Mizrachi and Sephardi Jews, whose family histories contradicted the simplistic graphics, were ignored. The aesthetic certainty of the infographic overcame the messy reality of Jewish identity every time.

By the end of 2020, it became clear that digital identity wars, besides excluding Jews, were trying to rewrite their history. And as tribes hardened, as influencers rose, and as social media became the primary arena for moral storytelling, Jewish identity found itself miscast in narratives too simplistic to encompass its ancient complexity.

When Identity Becomes Weaponized

Identity, once the cornerstone of liberation movements, became the terrain of tribal conflict. And for Jews, historically multilingual, multicultural, multi--ethnic, and multidimensional, there was simply no category clean enough to describe them. So, they became something else, the exception, the anomaly, the misunderstood outlier, and the perpetual hotspot in the matrix.

The identity wars set the stage for what came next, an online landscape primed for moral absolutism, allergic to nuance, and addicted to the clean geometry of oppressor and oppressed. When October 7 arrived, this machinery did exactly what it had been trained to do, *it collapsed complexity into instant judgment*. A generation fluent in infographics and tribal hashtags reached for the only frameworks it knew, even as the world convulsed with a horror too large for templates. Within minutes, the digital tribes hardened, narratives snapped into place, and misinformation began spreading faster than facts could be formed into a sentence. The identity frameworks that once claimed to explain everything suddenly proved incapable of explaining anything.

What followed was the most intense information breakdown of the social media era. And it began in the first 72 hours after the Hamas invasion.

Part 4

October 7 and the Digital Breakdown

Chapter 14

The First 72 Hours: Chaos, Misinformation & Narrative Collapse

The morning of October 7, 2023, began like any other Shabbat and holiday morning in Israel, quiet, peaceful, unhurried. By mid-morning, the world had changed. Hamas militants, along with some Gazan civilians, breached the Gaza border in a coordinated assault, murdering roughly twelve hundred people in Israel and kidnapping more than two hundred and fifty people back to Gaza. It was the kind of event that in any previous century would have been conveyed through radio addresses, delayed newspaper dispatches, and official briefings. But this was the age of instantaneous digital witnessing and the massacre unfolded not only in the fields and streets of southern Israel but across TikTok, X, WhatsApp, Telegram, Instagram, and encrypted channels, with the violence broadcast at the speed of notification pings.

The attacks were documented in real time by the attackers themselves. Hamas fighters wore body cameras and GoPros and livestreamed executions, uploaded videos through victims' phones, and circulated footage through their own Telegram channels. The world saw raw images before victim families were notified, before journalists reached the scene, and before governments could confirmed what was happening.

The deluge was overwhelming. Platforms that had spent years promising trust, safety and AI-driven content moderation collapsed instantly. Moderation teams were inundated with thousands of videos per minute. Automated classifiers, designed to distinguish graphic violence from news footage, failed utterly. Human moderators, many of them contractors working across continents without local knowledge, were unable to keep pace. Videos went viral faster than the platforms could label them, let alone remove them. TikTok's For You page turned into a factory-line of horrors. On X, algorithmic amplification pushed sensational footage to the top of the feed regardless of truth. Instagram Stories filled with screenshots, clips, and frantic reposts from users who perhaps believed they were helping, but often unknowingly shared terrorist propaganda.

For Jewish communities worldwide, the experience was surreal and traumatizing. The shock of the attack was immediate, but the digital trauma that followed was equally profound. Social media fed Jews throughout the world scenes of the massacre almost in synchrony with those living in Israel. Many saw videos showing people who looked like their cousins, colleagues, or grandparents being dragged from homes, shot in safe rooms, burned alive, or forced into vehicles at gunpoint. The horror had no time delay. It reached Jewish phones with the intimacy of a text message.

As the shock spread, the digital hostility began almost simultaneously. Within hours, diaspora Jews began receiving threatening messages and expressions of support for Hamas in their DMs and comments. These came not from extremist websites but from strangers and acquaintances on Instagram, TikTok, and X. In some cases, users who posted messages of solidarity with Israel found themselves doxxed by those who viewed any sympathy for Israeli victims as an act of aggression. It was a double trauma, the horror of the massacre itself, and the sudden realization that the new norms of the online world permitted Jewish victimhood to not only be unrecognized but openly mocked.

The digital infrastructure failed in five interconnected ways. First, there was *total information overload.* No platform in history had ever contended with such an immediate, massive influx of violent footage directly uploaded by perpetrators. The volume alone made content regulation impossible. Second, *the algorithms elevated the worst content.* Nothing is more intense, or more shareable, than graphic tragedy. Videos that should have been suppressed were instead pushed to millions. Third, *propaganda networks that had existed for years sprang to life instantly.* Accounts across the Middle East, Europe, and the U.S. began circulating doctored videos, mislabeled footage, and conspiracy laden narratives that reframed the events before the facts were even known. Fourth, *the internet fractured into instant framing battles.* Within minutes, hashtags expressing solidarity with Israeli victims competed against those asserting that the Hamas massacre represented legitimate Palestinian resistance (or did not occur at all). No narrative had time to establish itself and every narrative fought for dominance. Finally, *verification collapsed.* Credible journalists were drowned out by influencers, anonymous accounts, and emotionally charged clips whose authenticity no one could or cared to confirm.

The individual stories that emerged from those first hours were heart-breaking. Some families learned their relatives were kidnapped through a Hamas video circulated on Telegram. A father recognized his son in the back of a vehicle on his way to Gaza only because the clip showed up in his WhatsApp feed. Meanwhile, videos of atrocities supposedly committed by Israeli forces began circulating the very same day, except that many were taken from unrelated conflicts in Syria, Iraq, or Myanmar. One video of Syrian children gasping after a chemical attack was shared worldwide as supposed proof of Israeli brutality on October 7. Millions saw the video, but only a tiny fraction ever saw the correction.

Beyond the misinformation, there were bizarre and disturbing cultural reactions. In parts of the Arab world, TikTok users livestreamed videos of themselves watching the attack footage as if witnessing an event in a stadium. The distance that screens create between viewer and victim, already troubling in ordinary circumstances, became grotesque when applied to mass murder.

Platform responses were inconsistent and often too late. Meta attempted to block certain hashtags but did so sporadically, sometimes reinstating them accidentally. TikTok added Content Advisory warnings hours into the crisis, by which time millions had already seen unfiltered violence. Under Elon Musk, X had dismantled significant parts of its moderation infrastructure, allowing extremist accounts to regain visibility and misinformation to flourish. At one point, a fake account impersonating an Israeli Mossad spokesperson, @TheMossadIL, created online confusion and spread misinformation. The account had a significant following and used the official Mossad logo, leading many social media users and even some journalists to believe it was legitimate.

In the early hours, global sympathy overwhelmingly flowed toward Israel. World leaders issued statements and editorial pages expressed horror. The brutality was impossible to deny. But online, sympathy had a very short half-life. By the second day, misinformation, political ideology, and preexisting digital tribes began to reframe the massacre. Narratives emerged that depicted the attack as justified violence, while denying elements of the atrocity entirely. Some claimed the videos of the brutality inflicted on Israeli civilians were staged, and others argued that the victims were combatants rather than civilians. The shock waned, and ideology rushed in to fill the vacuum.

The speed of this narrative inversion was stunning. No major atrocity in modern history had ever had its narrative reframed so quickly. The world had barely processed the initial facts, and already the meaning of those facts was being pulled apart and reassembled in service of competing political identities.

The first 72 hours after October 7 revealed that the digital world has the capacity to reshape tragedy in real time. It accelerates it, reframes it, distorts it, and weaponizes it before the world has had the time to absorb what happened. The platforms designed for jokes, dances, and short form entertainment became the global delivery system for trauma, propaganda, and ideological warfare.

What followed in the days and weeks afterward wasn't just chaos but the emergence of a new, dangerous phenomenon—a digitally weaponized meaning-making machine that would reshape global understanding of October 7 long before history had a chance to intervene.

The Global Impact: How October 7 Hit Jewish Communities Around the World

The shock of October 7 ricocheted through Jewish life on every continent, riding fiber optic cables and algorithmic feeds faster than any journalist or statesman ever could. Within hours, Jewish communities in Paris, London, São Paulo, Sydney, Toronto, and Los Angeles were living inside the same catastrophe, but each through a different digital filter—French Telegram channels, British WhatsApp groups, Latin American TikTok, Australian news lags, and North American campus group chats. The massacre was the same, but the narratives that formed around it were not. In some places, empathy barely had time to appear before it was replaced by suspicion. In others, state rhetoric, local politics, and platform dynamics combined into a uniquely toxic mix. Understanding what the first 72 hours really did to the Jewish world requires a look at the shockwave as it broke, bent, and splintered across France, the UK, Latin America, Australia, Canada, and the United States.

France: From Shock to Panic to Abandonment

France, home to the largest Jewish population in Europe, felt the digital shockwave instantly. Within hours, violent misinformation circulated on French Telegram channels claiming Israel had staged the attacks. TikTok France pushed clips depicting the massacre as resistance. French far-left activists reposted Hamas propaganda nearly verbatim.

By day three, antisemitic incidents skyrocketed with reports of graffiti, threats, online harassment, and doxxing. Jewish parents removed mezuzah from their doorposts and Jewish students avoided campus. The October 7 digital narrative inverted so quickly in France that many Jews experienced a kind of emotional whiplash, with sympathy turning into suspicion in 48 hours.

United Kingdom: The Misinformation Blitz

In the UK, WhatsApp networks erupted. British Muslim influencers with huge TikTok audiences amplified unverified claims that Israel had fabricated the footage. Protests formed within hours, with some printed signs using the very images Hamas had produced. By day four, the London streets were packed, not in mourning for Jewish victims, but in jubilant, anti-Israel demonstrations.

Latin America: State Narratives and Algorithmic Poison

Latin American Jews experienced two forces, left-wing anti-Israel political framing, and algorithm-driven antisemitic content translated into Spanish and Portuguese. In Chile and Colombia, far left governments immediately reframed October 7 through a colonial narrative. In Mexico, TikTok misinformation, much of it from Middle Eastern accounts, became viral at a scale unmatched elsewhere in North America. Brazilian influencers pushed the false flag theory heavily. Jews in Argentina and Brazil reported WhatsApp disinformation clusters, especially strong in Sephardi and Mizrahi communities, where WhatsApp is the dominant communication tool. These private groups became overwhelmed by rapid--fire rumors, fake videos, and manipulated images.

Australia: Isolation Meets Global Chaos

Australia's Jewish community is small, tight-knit, and geographically distant from other Jewish communities, which caused Australian Jews to experience the trauma in a uniquely disorienting way. The massacre reached their smartphones in the evening. By the next morning, misinformation from the Middle East had already flooded TikTok Australia. Within days, anti-Israel protests filled Melbourne and Sydney.

Jewish parents described the surreal feeling of waking up to global chaos already in motion. Local media, slower and more cautious than American outlets, struggled to counter TikTok's immediacy.

Canada & the US: The Fastest Erosion of Moral Clarity

North America initially reacted with shock. But within 72 hours, university student groups, most notably at Harvard, issued statements blaming Israel and celebrating the massacre as resistance.

In Canada, misinformation spread through Telegram and TikTok, amplified by diaspora connections to Middle Eastern politics. Antisemitic attacks on Jewish communal institutions and Jewish owned businesses increased dramatically in Toronto, Montreal, New York, and Los Angeles, among other cities.

A Jewish Conclusion to a Very Digital Disaster

The first 72 hours after October 7 revealed what it means to experience atrocity in the algorithmic age. The digital world distorted the attack instantly, which collapsed the distinction between perpetrator and victim. It turned trauma into spectacle, and spectacle into propaganda. October 7 was the first Jewish catastrophe where millions of people watched in real time, and yet within hours determined the Jews were not the victims but oppressors who deserved their fate. More insidious than the ensuing chaos, what followed was the awakening of a new, dangerous phenomenon, *the algorithmic erasure of Jewish victimhood.*

The mechanics of that collapse and the way the digital world became the central battlefield of meaning, memory, and truth in the 21st century are discussed in the next chapter.

Chapter 15

The Victim-Villain Inversion: How the Jews Lost the Narrative in Record Time

The dust had not yet settled in southern Israel and families were still searching for loved ones. Bodies were still being identified as the world struggled to absorb the magnitude of what had happened on October 7. Yet, by October 9 and 10 the digital climate had shifted with astonishing speed. Barely forty-eight to seventy-two hours after the massacre sympathy that had initially flowed toward Israeli victims began to reverse. Confusion morphed into skepticism, skepticism hardened into denial, and denial metastasized into an accusatory certainty that Israel, not Hamas, bore responsibility for the violence.

In the span of two days, the digital public square managed to invert the victim and the villain in the fastest reversal of moral clarity of the modern information age. The platforms not only failed to hold the narrative, *they actively rewired it.*

How this happened is not a simple question emotionally. Nor is it politically simple. But one thing is undeniably true; it was technologically simple. The reversal fit the logic of the internet so seamlessly that it almost felt preprogrammed.

The Explosion of Misinformation

By October 9, viral posts claimed that Hamas had not committed atrocities at all. Videos that clearly showed kidnappings, murders, and burnings were dismissed as fabrications, deepfakes, or IDF propaganda. Within forty-eight hours, millions of users had consumed posts insisting that the hostages were actors, or that the massacre was exaggerated, or that Israelis had fabricated evidence to justify retaliation. These narratives spread faster than reputable journalism because they obeyed the rules of virality in that they were fast, shocking, emotional, polarizing, and easy to share.

Influencers with enormous audiences began amplifying misinformation without verification. Some had tens of millions of followers. Their posts, filled with unsubstantiated claims, traveled farther and faster than any correction could. Days later, some posted apologies, but by that point, the damage had been done. The narrative they helped shape had already calcified. The victim–villain inversion was not a slow process; *it was algorithmic.*

Why the Narrative Inverted So Fast

The speed was shocking, but the mechanics were predictable. A perfect storm of technological, cultural, and ideological factors converged. Why did this happen?

Weak Media Literacy Among Younger Users

Most Gen Z users get their news not from newspapers, TV, or long form journalism, but from TikTok, Instagram, and X. These platforms favor brevity over accuracy. The average political TikTok video is under thirty seconds. The average fact check takes between minutes and hours to produce. Credibility isn't a concern. What matters is speed and simplicity.

In this environment, a blurry clip uploaded by a random account often carried the same weight as one verified by a reputable news organization. For millions of users, if a video looked compelling, it was true. And if it aligned with their political framework, it *felt* true.

Algorithmic Preference for Emotional Victimhood

Platforms reward content that portrays suffering. Posts depicting Palestinians as victims triggered powerful emotional responses and consequently were boosted by algorithms. This is not intentional bias, yet it is bias by design. The system responds to engagement patterns, not facts.

Meanwhile, content showing Israeli suffering was quickly overcome by the far greater volume of political content reframing the massacre. Videos of murdered families could not compete with the narrative machinery already in motion.

Pre-existing Ideological Frameworks

For years, a simplified ideological map had dominated activism on social media matching oppressor vs. oppressed, white vs. non-white, and colonizer vs. colonized. Long before October 7, Israel had been slotted neatly, and inaccurately, into the oppressor/colonizer category. Jews, particularly Ashkenazi Jews, were widely described as white, regardless of history, genetics, or diaspora realities. Mizrachi and Sephardi Jews were erased almost entirely.

So, when Hamas committed atrocities, many online activists instinctively rejected the possibility that Palestinians, perceived as the oppressed, could also commit acts of savagery. Their ideological framework did not permit it. Instead, they sought explanations that preserved their worldview, such as the massacre must be exaggerated, Israel must be lying, the victims must not be innocent, or the footage must be manipulated. Prejudices caused the reversal. Pre-existing frameworks empowered it.

The Influencer Economy

Influencers have no editorial standards. Regardless of their sincerity, their incentive is engagement and audience retention. Posting an unpopular stance risks losing followers, while posting a popular one generates likes, comments, and shares. On October 8 and 9, the popular stance across many online communities was to frame the attack as part of a liberation struggle. Influencers chose virality over truth. They simply posted what their followers expected to see. An entire moral universe was shaped not by philosophers or clerics, but by content creators afraid of losing engagement.

Coordinated Disinformation Networks

Behind the chaos lay intentional actors. Iranian, Qatari, and Russian disinformation networks, already active for years, sprang into action. They flooded platforms with anti-Israel narratives, fake videos, distorted quotes, and inflammatory memes. Some activist networks, both far left and far right, joined the amplification for their own ideological purposes. Anonymous meme accounts, operating through Telegram and X, produced staggering amounts of content.

The victim-villain inversion did not emerge organically. It was pushed, relentlessly, by machinery designed precisely to collapse narratives.

The Jewish Experience: Betrayal in Real Time

For Jews, the narrative collapse was not abstract. It was personal, and it was devastating. Jewish professionals opened social media posts expecting sympathy or at least silence, only to find colleagues reposting claims that Israeli victims were liars, actors, or colonizers who got what they deserved. Jewish college students watched classmates, including peers they had studied with, sign open letters blaming Israel entirely for the events.

Jewish influencers were unfollowed by friends and strangers alike for expressing grief. Some were bullied into staying silent because posting about murdered Israelis was seen as politically insensitive. Others were pressured to balance their mourning with statements condemning Israel. In identity spaces that prided themselves on empathy, Jews were told their grief was unaligned with the movement, their fear was inconvenient, and their trauma was suspect.

Stories from the Front Lines of Narrative Collapse

The speed and severity of the reversal revealed itself in stories that spread through Jewish communities like aftershocks.

At Stanford, Harvard, and other elite universities, student groups released statements, some within hours of the attack, declaring that Israel was entirely at fault. Many had not yet seen a single verified report. Some did not even acknowledge the massacre at all. Jewish students were stunned, then terrified, as rallies praising Hamas appeared before vigils for the victims could even be organized.

In dozens of cities, pro-Hamas rallies formed in the physical world after they formed online. Digital enthusiasm became real world action. Mobilization began with viral misinformation, then moved into the streets.

Jewish public figures, actors, musicians, and writers, were publicly shamed for mourning Israeli dead without simultaneously condemning Israeli policy. The mob took control and would not allow Jewish trauma to exist, unless it was politically approved.

The Psychological Shock: Losing the Narrative Before Grief Could Form

Historic massacres typically generate sympathy for victims. That empathy might fade over time, but it is rarely reversed within days. The inversion that followed October 7 was historically unprecedented.

Jewish communities had barely begun mourning when the global digital chorus turned accusatory. The online world took a single look at Jewish suffering and declared that the Jewish narrative didn't fit its prevailing ideology. For many Jews, this confirmed the long feared truth that digital identity politics had no space for Jewish existence outside predetermined roles. More than painful, the speed of the reversal was disorienting. It stripped Jewish communities of the ability to process grief before forcing them to defend the legitimacy of that grief.

Historical Parallels in the Digital Age

The inversion echoed older patterns. During the Black Death, Jews were massacred by Europeans who accused them of poisoning wells, despite the fact that Jews were equally devastated by the plague. After medieval expulsions, Jews were blamed for the economic decline that followed their departure. After pogroms, survivors were often charged with provoking violence. History's cruelest pattern is the inversion of blame onto Jewish victims.

The internet accelerated that pattern to an extraordinary extent. What once took months or years to calcify now took days. The mechanism by which it did so was no longer rumor or superstition but an algorithm.

Sympathy Was Brief. The Battle for Meaning Was Immediate

The massacre of October 7 should have been a moment of moral clarity. Instead, it became the opening act of a narrative war in which the truth was drowned in ideology, algorithmic incentives, and coordinated disinformation campaigns.

Jews lost the narrative not because their story was weak, but because the digital world had been primed to reject it. The identity frameworks, activist ecosystems, and influencer economies of the last decade collided with the worst atrocity in modern Jewish history, and the result was a narrative collapse unprecedented in scope and speed.

The moral inversion was only the beginning. In the days that followed, this narrative war would escalate into something even more corrosive, a digital battlefield where misinformation hardened into doctrine. This fact raises the most uncomfortable question of all, *would such an inversion of morals have taken place if the victims weren't Jewish?*

Chapter 16

Could the Inversion Have Happened If the Victims Weren't Jewish?

The internet performs moral gymnastics at gold-medal levels, but only when Jews are involved. In the days after October 7, as the world distorted what had happened with a mixture of righteousness and mockery, a troubling question rose from Jewish communities across the globe—spoken at vigils, shouted in group chats, and asked in classrooms, boardrooms and synagogue parking lots—*Could this have happened if the victims weren't Jewish?*

Could the world have watched videos of teenagers being hunted, parents being executed in front of their children, elderly women being dragged across borders, and within forty-eight hours begin insisting the massacre was exaggerated, fabricated, or deserved? Would the narrative have inverted so fast if the bodies had belonged to another people?

The question was not rhetorical. It was a genuine attempt to make sense of the inversion itself, how global sympathy evaporated in less than three days, how denial began before mourning, how justification outpaced facts, and how images of agony were met with suspicion not solidarity. As painful as it was to ask, the answer was even more painful. *No. Not like this. Not this quickly. Not with this confidence. Not with this ferocity.*

The speed of the inversion was not merely a byproduct of online chaos. It was a function of something older, deeper, and more structural, something baked into the cultural, ideological, and algorithmic machinery long before October 7 ever happened.

To understand the inversion, we must understand that Jewish victimhood occupies an unusually fragile position in the modern imagination. And to understand that fragility, we must examine the forces, technological and ideological, that collided with the oldest hatred on earth to produce a reaction unlike anything the digital world had ever seen.

The Digital Reflex to Distrust Jewish Suffering

There is a long, grim history of Jewish suffering being denied, minimized, or rationalized. The psychological pattern is ancient, when Jews suffer, many societies search for reasons the suffering is deserved. This reflex, instead of disappearing in the digital age, was amplified.

After October 7, Jewish trauma was subjected to an extraordinary level of scrutiny. Videos of murdered families were dissected by strangers on TikTok as if they were forensic scientists looking to solve a stubborn cold case. Grief was interrogated as political messaging and hostage stories were treated as propaganda. Survivors were accused of lying and journalists were smeared for reporting what they had seen with their own eyes.

Jewish communities found themselves forced into a defensive posture almost immediately. They were not allowed to grieve. Instead, they were required to explain and justify. And when they did speak, their voices were drowned out by accounts insisting that Jewish suffering was fabricated, exaggerated, or weaponized.

What was at work here was not just disbelief, but resentment that Jewish vulnerability existed at all. This dynamic is not random, but rooted in antisemitism not in the overt, swastika waving sense, but in the subtler assumption that Jewish suffering need not be acknowledged because Jewish power is always imagined as excessive.

The conflation of Jews with power, money and influence and centuries--old tropes fed directly into the speed of the denial. If Jews are always seen as powerful, then by definition they cannot be victims and so any claim of victimhood is suspect. The internet took this logic and scaled it globally within forty-eight hours.

The Ideological Priming That Made Inversion Inevitable

The online Left had spent years framing Israel exclusively as a colonial aggressor. The online Right had spent years flirting with conspiracies about

Jewish control. Foreign adversaries had spent years building disinformation pipelines targeting Jewish and Israeli institutions. Influencer culture had spent years treating nuance as the enemy of engagement. Algorithms had spent years rewarding trauma content, especially when attached to certain narratives.

So, when October 7 happened, the digital ecosystem did not need to invent a narrative from scratch. It already had one locked, loaded, and ready to deploy. This identity template collapsed thousands of years of Jewish history into a single, misleading contemporary framework. And once the template was in place, the idea of Jewish victimhood simply would not compute.

Had the victims been another group, Black, Indigenous, Muslim, or immigrant, the inversion would have either taken far longer or would not have occurred at all. It would have required too much cognitive dissonance. But Jewish suffering already came pre-attached with a conceptual loophole, the belief that Jewish pain is never purely pain but always political.

This is how antisemitism creates the conditions for inversion before the event even occurs. The denial was not a reaction to October 7, but the fulfillment of an ideological script written long beforehand.

The Fragility of Empathy for Jews in the Online Imagination

Empathy for Jews is often strangely fragile, generous at first, but brittle under pressure. The world was briefly horrified by the massacre. But that empathy had a fast expiration date. It took but two days and the digital tide shifted not because facts had changed, but because users reverted to frameworks more comfortable than the reality confronting them.

If the victims had not been Jewish, the empathy window would likely have remained open much longer. If the attackers had not been Hamas, the political stakes might have been lower. If the narratives around Israel had not been preloaded into the activist and influencer economy, the discourse might have remained grounded in basic morality for more than seventy-two hours. But because the victims were Jewish, empathy was fleeting, conditional, fragile. This fragility is itself a form of antisemitism, not hateful, but structural: a prejudice woven into the ideological and emotional algorithms of the digital age.

Antisemitism as the Accelerant

Antisemitism determined the speed at which the narrative collapsed. It determined which claims went viral, and which videos were believed, determined whose grief required justification and whose humanity was to be questioned. Remove antisemitism from the equation, and the inversion still might have occurred, slowly, carefully, awkwardly. With antisemitism in place, the inversion was instantaneous, confident, and global.

October 7 Was Not Only a Terrorist Attack. It Was a Stress Test of How the World Sees Jews

If the massacre revealed the fragility of Jewish safety, the narrative collapse revealed the fragility of pro-Jewish empathy. The question, Could this have happened if the victims weren't Jewish?, is not about comparing suffering. It is about understanding why Jewish suffering becomes uniquely vulnerable to erasure, denial, or inversion. The world showed, in real time, that Jewish grief is negotiable, Jewish trauma is disputable, and Jewish victimhood is reversible in a way that few other groups experience. The digital age did not create that dynamic, but it exposed it with ruthless efficiency.

As the aftermath of October 7 continues, that exposure becomes central for understanding the crisis, as well as the world Jews must navigate in its wake. Because what happened online in the days after October 7 was not just a glitch in the system. *It was the system.*

Chapter 17

Digital Warfare: Propaganda, Bots & the Attention Battlefield

In the weeks following October 7, it became unmistakably clear that the war was not confined to Gaza, Israel, or the Middle East. It wasn't contained within borders, terrain, or even the physical world. A second war raged simultaneously, fought not with rifles and rockets but with videos, hashtags, hacked accounts, psychological pressure, and algorithmic manipulation.

This was the first modern conflict in which digital warfare was not an accessory to the battlefield, but *part* of the battlefield. It played out across screens worldwide, shaping global perception in real time. Governments, terror organizations, state actors, non-state activists, influencers, celebrities, bots, and ordinary users all took part. The fight was for attention, belief, and emotional dominance. The weapons were livestreams, disinformation campaigns, deepfakes, and algorithmic acceleration.

For Jews around the world, who were both the targets and the observers of this information war, the experience felt like data bombardment. The goal wasn't simply to influence what Jews thought, but to weaken their sense of safety, fracture their identity, and destabilize their emotional resilience. The digital war had consequences, and it hit Jews everywhere.

The New Battlefield: The Internet as National Security Theater

Before October 7, major militaries, including Israel's, already recognized the internet as a domain of warfare on an equal footing with land, sea, air, and space. The U.S. had cyber commands, Israel had Unit 8200, the Russians had their troll farms, and Iran had its cyber brigades. Hamas, in a game changing move, had TikTok channels capable of distributing psychological warfare to millions with little more than a smartphone.

On October 7, as massacres unfolded on the ground, Hamas simultaneously executed a digital offensive with brutal sophistication. Hamas members strategically uploaded bodycam footage, hostage videos, and torture clips, with the intent to terrorize not only Israelis, but Jews everywhere, to unsettle diaspora communities, to spark outrage among global audiences sympathetic to Palestinians, and to overwhelm platforms before moderators could react.

The videos were graphic and intentionally cruel, psychological warfare disguised as documentation. Meanwhile, the Israeli Defense Forces launched their own digital counteroffensive with rapid briefings, translated updates, annotated maps, drone footage, and real time fact corrections. Their social media accounts became emergency broadcasting centers. The battle wasn't just about who is winning on the ground, but also about who controls the narrative.

Yet the most complex digital activity wasn't coming from the military actors themselves. It was coming from the invisible armies of bots, troll farms, coordinated networks, hacked accounts, and state funded disinformation machines. This was the attention battlefield, and it was as consequential as the physical one.

The Weaponization of Attention

Digital warfare operates on the simple principle that you don't need to win the argument; you only need to overload it. One major tactic is *attention saturation*, the flooding of the internet with so much content, videos, memes, lies, and half-truths, that truth becomes impossible to distinguish from noise. Even credible journalists lose their footing. Even allies become confused.

During the first days after October 7, bot networks pumped out thousands of posts per hour under every major conflict-related hashtag. Their goal was to drown authentic reporting in a tsunami of unverifiable material. By the time fact checkers mobilized, the algorithms had already elevated the emotional, the graphic, and the inflammatory. Narratives became weapons.

So did slogans and hashtags. It didn't matter whether the material was true, it mattered that it was viral.

Narrative Flooding: The Repetition That Rewrites Reality

Another tactic of digital warfare is *narrative flooding*. The idea is to repeat a small set of emotionally powerful claims until they dominate the conversation. After October 7, certain words and phrases began appearing everywhere online, amplified by hostile actors. They included terms like genocide, apartheid state, colonizer, ethnic cleansing, starvation, war crimes, and resistance by any means, to name but a few.

Some of these terms had academic origins, while others had activist origins, but their online spread was not accidental. State and non-state disinformation networks deliberately poured these terms into the digital bloodstream until they became the default framing for millions of users. Once a narrative template is established, even contradictory evidence is forced to bend around it.

This is a type of conditioning. And once conditioned, users often react emotionally, on the basis of the narrative they've already absorbed and not on facts or evidence. This is the battleground of perception.

Hashtag Hijacking and the Bot Armies in the Shadows

One of the most visible components of digital warfare consists of *hashtag hijacking* or *flooding hashtags* used by one group to outshout the other through overwhelming volume. Within hours of October 7, hashtags like #IsraelUnderAttack, #HamasAttack, and #BringThemHome were overwhelmed by bot farms posting propaganda, manipulated videos, political slogans, or anti-Israel memes. Entire campaigns vanished beneath a flood of coordinated posts.

Researchers later traced these floods to bot farms linked to Iran's IRGC cyber operations and, in some cases, to state adjacent Russian troll networks. Their strategy was to suffocate. By drowning out messages posted by Israelis and Jews around the world, they hijacked the debate and made their narrative the only story line. Far from being an expression of grassroots or spontaneous reactions, this situation was engineered.

Manipulating Sympathy: Emotional Warfare at Scale

In any conflict, images of suffering people, especially women and children, carry enormous emotional power. In digital warfare, such images can be deliberately weaponized. After October 7, graphic images of civilian casualties in Gaza spread widely across the internet. Many were real, but just as many were taken from conflicts in Syria, Iraq, Libya, or Afghanistan and falsely labeled as coming from Gaza. Some images were manipulated and others were edited to remove identifying backgrounds or timestamp discrepancies. The aim was to elicit shock, grief, disgust, and rage. Accuracy was a secondary consideration, if a consideration at all.

This tactic, sometimes called *sympathy manipulation*, works because social platforms boost emotional content. When users scroll through a feed and encounter image after image of anguished civilians, their moral and political reactions develop faster than their fact checking instincts.

Meanwhile, footage of Hamas atrocities, authentic, verified, and time--stamped, was increasingly labeled as fake, AI generated, or unconfirmed. Some users, having consumed hours of civilian suffering in Gaza, refused to believe Israeli victims existed at all. This was an indication of the success of an emotional propaganda cycle.

Deepfakes and Fake Officials: The New Tools of Deception

The appearance of AI-generated content marks a disturbing evolution in digital warfare. Within days of the attack, deepfake videos began circulating purporting to show Israeli officials admitting to war crimes, IDF soldiers confessing to atrocities, and fake leaked recordings of Israeli leaders discussing fabricated plans to starve or slaughter Gazans. Other deepfakes include a fabricated UN spokesperson video which circulated for days before being debunked, and fake Red Cross statements taken as authentic. None of these were real, and yet all were widely shared.

As AI tools improve, these tactics will only proliferate. The line between truth and fiction will continue to erode. In traditional warfare, armies try to destroy infrastructure. In digital warfare, the target is the public's ability to believe its own eyes.

The Jewish Experience: A State of Digital Siege

For Jews worldwide, the digital battlefield became a psychological front line. Each day brought new disinformation designed to demoralize, frighten, outrage, or isolate them. Jewish institutions, synagogues, schools, and federations implemented enhanced cybersecurity protocols. Some were hit with ransomware—malicious software which encrypts files, systems, or data, making them inaccessible until a ransom is paid. Others faced waves of bot driven harassment. A few were targeted in attempts to breach internal communications. Community leaders found themselves thrust into the unfamiliar role of digital crisis managers. For the first time in memory, Jewish communities globally were navigating an environment in which their psychological safety was under constant assault.

From Pogrom Rumors to Telegram Lie Factories: The Long Arc of Jewish Disinformation Threats

What Jews experienced after October 7 was new in speed, scale, and technology, but not in pattern. Historically, Jewish safety has always been vulnerable to the velocity of rumor. In late 18th century Eastern Europe, a single whisper of a blood libel could travel from the tavern to the whole town in a day and was often enough to spark a pogrom. Antisemitic disinformation spread slowly, almost quaintly by modern standards, but its impact was immediate and lethal. Jews were rarely attacked for what they did, but for what someone said they did.

Digital platforms supercharge this dynamic. The same mechanisms that once spread blood libels now spread doctored videos. The same societal vulnerability to fear and suspicion that once incited mobs now fuels algorithm-driven outrage. What has changed is not people and the hatreds they hold but the machinery that carries them.

The violence that followed October 7 unfolded at the speed of fiber optic cable rather than rumor. But at its core, it followed a pattern Jews have confronted for a thousand years; *falsehood leads to emotion, emotion leads to belief, and belief leads to violence.*

Lessons from the Digital Battlefield

The lessons of this era of digital warfare are sobering. Propaganda now travels farther and faster than truth. Bots can mimic crowds, and deepfakes can mimic humans. Hashtags dictate narratives and algorithms reward the most extreme emotions. Misinformation has become a weapon.

The digital battlefield has no rules, Geneva Convention provisions, ceasefires, or borders. It is always on, always expanding, and always adapting. On this battleground the weapons are images, words, code, and algorithms. The violence of October 7 did not end with the physical attack, but rather it continued in the digital realm, reshaping perceptions, isolating victims, and destabilizing communities far from the physical battlefield. The internet became a place where accuracy was optional, but outrage was mandatory.

Perhaps the hardest truth is that the digital war is still ongoing. But there are some things we can be doing.

Digital Defense Begins with Literacy, Not Slogans

The first and most enduring form of Jewish digital defense is media literacy, the ability to recognize manipulated videos, spot coordinated bot activity, distinguish emotional propaganda from verified reporting, and understand when a narrative is being manufactured for psychological impact. This isn't about political alignment, but about knowing how the digital ecosystem works. In a world where deepfakes circulate before facts and where bots outnumber humans, literacy becomes a form of armor. Jewish teens, parents, educators, and communal leaders can all benefit from understanding how attention algorithms operate, how disinformation spreads, and how to evaluate sources before sharing or reacting. In a digital war safety is not in silence, but in discernment.

Community Infrastructure is its Own Shield

Jewish institutions, schools, synagogues, federations, and advocacy groups must strengthen digital resilience in the same way they've strengthened physical security. That includes cybersecurity protocols, rapid response communication strategies, partnerships with tech literate experts, and mental health support programs for members who experience online harassment. These are communal survival tools. When a synagogue's Instagram posts are targeted by coordinated trolls, or when a Jewish high schooler receives a flood of abusive DMs, the most valuable defense isn't a counterargument, it's a support system that knows how to intervene, document abuse, report it effectively, and care for the people affected by it.

Solidarity and Storytelling Remain Ancient Tools With Modern Power

Jews must also defend themselves in the digital war by refusing erasure. That means telling personal, historical, and spiritual Jewish stories in ways that affirm identity rather than argue politics. It means elevating Jewish diversity with Mizrahi and Ethiopian stories, queer Jewish voices, and secular and observant experiences alike. Propaganda thrives where people are reduced to caricatures, but it falters when confronted with humanity. The digital battlefield punishes complexity, and Jewish life is nothing if not complex. We need to simplify it through our very human experiences. Choosing to be visible, nuanced, and authentically Jewish is a way of occupying digital space without surrendering to its distortions.

Can the Digital War be Won or Only Survived?

Whether digital war can be won is uncertain. The battlefield shifts too quickly, the technologies evolve too rapidly, and the attention economy rewards the very forces that distort reality. And there's also that pesky antisemitism/antizionism to contend with.

But it can be survived, and survival has always been the most rebellious part of the Jewish story. Victory may not mean defeating disinformation outright. It may instead mean building enough communal resilience, enough literacy, solidarity, and emotional stability, to withstand the waves when they come. The digital realm may never be fully safe, but Jews have endured far more dangerous terrains. The goal is not to conquer the algorithm. The goal is to remain Jewish within it.

Chapter 18

The Media Feedback Loop: How News and Social Platforms Reinforced Each Other

Traditional media and social media now operate as a single, mutually reinforcing ecosystem. One is slow and institutional, the other is fast and emotional, and antisemitism flows through the seams where they meet. Their intersection not only amplifies hate, but accelerates it, legitimizes it, and gives it new pathways to reach the mainstream. The two systems interact, collide, and contaminate each other.

Traditional Media Give Fringe Narratives Social Credibility

Antisemitic narratives often emerge on anonymous platforms like Telegram or 4chan.They are fringe, until a columnist, or TV pundit, or headline refers to them. Even critical coverage inadvertently lifts these ideas out of the gutter and into the realm of *Things Reasonable People Are Discussing.* As an example, a trope about the Rothschilds controlling global finance saw a second life not because journalists endorsed it, but because they reported on how widely it was spreading. The moment traditional media name a conspiracy, it becomes mainstream enough to discuss.

Antisemitic Incidents Go Viral Before Facts Exist

Social media demand instant reaction, while traditional media supposedly insist on confirmation. Those two speeds are incompatible. Social media respond in seconds, spread unverified claims, reward emotional content, punish nuance and are dominated by video, which is easy to mislabel. Traditional media respond in hours, require verification, follow (or are supposed to follow) editorial standards, and typically avoid showing graphic violence.

So, when antisemitic/antizionist flashpoints occur, like attacks, protests, or flare ups in Israel, the first version of the story is almost always social media's version. Once a false early narrative gains traction, traditional media often adopt its frame even when debunking it. Emotion beats accuracy when accuracy arrives too late.

Traditional Media Give Antisemites/Antizionists Institutional Footing, Social Media Give Them Scale

Historically, antisemitic ideas gained power when endorsed by institutions like newspapers which published blood libels, state radio which broadcasted propaganda, or intellectuals who endorsed pseudoscience.

Now, social media provide a new multiplier that bypasses the institution and provides independent virality. When traditional media pick up a viral antisemitic claim, even to contextualize or discredit it, they lend institutional weight that extremist networks then cite as proof of legitimacy. Influence flows upward from social media to traditional media, and then downward from traditional media to social media, where it is used as validation.

Whatever traditional media publish becomes cropped, reframed, memed, stripped of context, and repurposed for propaganda. Extremists rely on this process. They aim to coax a mainstream headline, so that they can clip it, spin it, and flood it into digital communities.

Social Media Permeate the Editorial Board

Journalists live online. Almost everything flows through Twitter/X, Instagram, and TikTok. their sources, their background information, their personal biases all emanate from their online world. This shapes how stories are framed.

Some reporters begin subconsciously adjusting coverage to avoid backlash or follow online trends. Some customize reporting to signal tribal alignment or protect their social reputation. Through them antisemitic/antizionist tropes slip into respectable outlets, not necessarily because editors endorse them, but because digital pressure shapes editorial judgment. The algorithm becomes an invisible editor.

Further contaminating editorial positions is the publish-it-anyway end-run reporters sometimes employ when they believe an editor's call is wrong (too cautious, too political, too legalistic, too advertiser-sensitive) and bypass the internal chain of command and push the piece (or a substantial excerpt) directly to the public on social media, often with the pointed subtext that this is what the newsroom tried to suppress. In a media environment where individual journalists cultivate direct audiences and personal brands, this kind of public disclosure can feel less like insubordination and more like accountability, especially when the journalist frames the editor's intervention as suppression rather than standard editorial judgment.

This tactic comes with real costs, even when it's motivated by principle. It turns editorial deliberation, normally a private process designed to improve accuracy, fairness, and legal safety, into public combat, and it pressures editors to defend decisions in the same reactive arena they're trying to keep away from. Over time, it weakens the authority of editing as an institution, incentivizes reporters to treat gatekeeping as a reputational opportunity, and invites audiences to assume that anything unpublished must have been squashed for the wrong reasons. This phenomenon reflects a deep shift in journalism, which is increasingly built around direct distribution and personal trust, while newsrooms are still built around collective standards.

Simple Narratives from Social Media Outpace Traditional Journalism

We have reviewed how social media are persuasive because they offer simple narratives of oppressor vs. oppressed, predator vs. victim, and colonizer vs. colonized, and how such narratives fit perfectly into antisemitic tropes, especially those which are anti-Israel. Thus we can explain, at least in part, the post October 7 meltdown as emanating from a pre-loaded library of moral templates. Once these templates dominated the digital atmosphere traditional media end up absorbing them, sometimes out of bias, but mostly because editors and reporters operate inside the same attention ecosystem as everyone else. They see what is trending, what is being shared, and what language is being used, and they begin to mirror it.

It is not that the news organizations consciously chose a narrative, but that the narrative had already been chosen for them by the algorithmic winds shaping public opinion. Social media write the script, and when it comes to Jews or Israel scripts often draw on false, antisemitic narratives.

Disinformation Manufacturing: A Symbiosis

What makes the modern information ecosystem so dangerous is not only the presence of disinformation, but the *symbiotic loop* that forms between bad actors, social media platforms, and mainstream journalism. State actors, extremist groups, and organized propaganda networks understand this loop much better than democracies do. They know that if they inject a toxic idea into social media, it only needs to gain traction, and the traditional media will stamp it with credibility. And a few thousand bot accounts can manufacture traction within minutes.

Once a narrative or phrase begins trending, journalists see it rising on their dashboards and in their analytics feeds. The temptation to join in on the trend, even in order to discredit its content, is too often too hard to resist. The coverage granted the trending topic increases interest, which in turn increases online searches on the matter. Social platforms detect this increased interest, and their algorithms begin surfacing more of the same kind of content. Suddenly, what began as a fringe narrative whispered in extremist corners bursts into the mainstream. Journalists then cover the backlash to the original backlash. The result is that an entire nation finds itself arguing over a propaganda product designed to destabilize it.

This is how digital antisemitism/antizionism evolves, not as a linear message but as an industrial ecosystem. The commercialization of news in the West has opened the door to changes in editorial priorities that pair the interests of traditional media with the whims of social media.

The Emotional Economy Rewards Antisemitism Everywhere

Traditional media reap rewards from conflict, bold claims, and polarizing commentary. Newsrooms are built on the theory that drama keeps audiences watching. Social media, by contrast, are rewarded by the more primal adrenaline rush of choosing sides. They more than just tolerate emotional extremity, they actually require it. A post that calmly explains a certain perspective sinks to the bottom of the feed like a stone thrown into the ocean, while a post that stokes identity or moral panic rockets to the top as if it's strapped to a SpaceX booster.

Antisemitism/antizionism thrives in both ecosystems because it fits neatly into each incentive model. For traditional media antisemitic narratives provide a simple antagonist for otherwise messy geopolitical stories. They can frame Jews (or Israel) as the central characters (usually villains) in conflicts that are too complicated to explain in the allotted 800 words or 3-minute broadcast segment. For social media, antisemitism is algorithmic jet fuel. It condenses complex anxieties into easily digestible memes, points to a recognizable scapegoat, and reinforces group identity with lightning efficiency. It gives users the emotional payoff of certainty in a world defined by ambiguity.

Most importantly, antisemitism functions like a universal narrative adapter, sliding effortlessly into the story lines of influencers, whether left wing, right wing, libertarian, or nationalist. It plugs into economic fears, racial anxieties, anti-government sentiments, anti-colonial rhetoric, and conspiracy networks without needing to be reinvented each time. Its flexibility is its danger since it can be wrapped in academic jargon, populist anger, religious conviction, or meme culture.

That is why Jews so often find themselves in the eye of the storm when global narratives fracture. Antisemitism satisfies the media's hunger for conflict and social media's appetite for outrage. In the dual economies of traditional and digital discourse, that makes it endlessly, tragically, and virally useful.

How Traditional Media and Social Media Intersected in the Post-October 7 Global Response

The aftermath of October 7 revealed a new information ecosystem, one in which traditional media and social media no longer operate as separate domains. Instead, they form a single, fused, mutually reinforcing system, where narratives travel back and forth in minutes, distortions jump across platforms instantly, and legitimacy and virality feed each other in a continuous loop.

What emerged after the massacre was a nightmarish fusion of institutional authority and algorithmic chaos, in which social media set the emotional frame and traditional media willfully inherited that frame. Social media weaponized the coverage and traditional media reinforced the weaponization.

For Jews watching this unfold was disorienting. Suddenly a world where the most trusted institutions and the least reliable influencers were speaking the same emotional language that intentionally disregarded Jewish trauma, complexity, and truth.

Social Media Created the Initial Frame Before Newsrooms Even Woke Up

On October 7, TikTok, Telegram, Instagram, and X, became the world's first reporters—not journalists. The footage uploaded by Hamas spread the resistance narratives instantly. Traditional media, from the BBC to CNN to The Guardian, woke up to a world whose emotional judgments were already set by social media. This narrative suited their own perverse world view, and they did little, if anything, to call out, contradict, or counter the prevailing moral inversion.

A look at a broad example of Western media's coverage of the Israel--Hamas conflict post October 7 shows a pursuit of so-called balance had come at the expense of truth, and at a moment when accuracy was a matter of global consequence.

The BBC

The BBC's coverage of the post-October 7 war quickly became a case study in how legacy media, under pressure from both digital virality and long-standing editorial instincts, can frame a conflict in ways that distort rather than clarify. In the immediate aftermath of the massacre, the BBC struggled to describe Hamas members as terrorists, clinging instead to euphemisms such as militants or fighters, even as the organization livestreamed the kidnapping, execution, and mutilation of civilians. While the BBC claims neutrality, this linguistic choice is not neutral. It elevates a murderous ideology to the status of a political faction and flattens the moral landscape at precisely the moment clarity is needed. Jewish communities worldwide watched in disbelief as the broadcaster treated the intentional mass killing of Jews as a morally ambiguous event, a framing that set the tone for much of its later reporting.

As the days unfolded, the BBC increasingly relied on claims of Gaza casualties issued by the Hamas-run Gaza Health Ministry, figures released without transparency, verification, or independent access, yet presented with the authority of established fact. This practice, repeated across broadcasts and BBC push alerts, created a perception of Israeli aggression divorced from context or nuance and gave Hamas extraordinary agenda-setting power. When Israeli officials disputed claims or requested corrections, those objections were framed as partisan pushback rather than legitimate challenges to unreliable data. The result was a feedback loop in which the BBC amplified the narratives circulating most intensely online, particularly those that fit pre-existing assumptions of asymmetry and colonial framing, even when the underlying evidence was thin or contested.

Perhaps the most damaging aspect of the BBC's coverage was its failure to interrogate the digital propaganda war unfolding in parallel with the conflict. The TikTok narratives, doctored videos, bot-amplified hashtags, and AI-generated images were reported by the BBC without scrutiny regarding their origins or their strategic intent. When false stories spread, the BBC repeated the allegations before verifying them, contributing to an explosion of global protests and, in some cases, real world violence against Jews. Even after evidence overturned the initial claim, the correction lacked the urgency and visibility of the original error. In this way, the BBC not only reflected public sentiment, but it became entangled in the very disinformation ecosystem it was supposed to counter.

A review of the BBC's official Corrections and Clarifications page, as of early 2025, showed nine corrections retracting false or misleading information about Israel. The BBC's Arabic service has had to make significantly more corrections. According to a report by media watchdog CAMERA (Committee for Accuracy in Middle East Reporting and Analysis) 216 corrections and clarifications have been made in connection to BBC Arabic content since October 7, 2023, for stories deemed biased, inaccurate, or misleading. This averages about two stories a week.

CNN

CNN's coverage of the Israel-Hamas war revealed a newsroom struggling to reconcile its breaking-news instincts with the unprecedented speed and manipulation of the digital information environment. In the first hours and days, CNN aired raw, unverified footage circulating on social platforms, often without sufficiently emphasizing its origins as Hamas Telegram channels, and thereby giving global broadcast legitimacy to material designed as psychological warfare. The network's commitment to immediacy frequently outpaced its commitment to verification. While CNN has a long tradition of live reporting from conflict zones, this conflict exposed a vulnerability: when speed becomes the primary value, groups skilled in digital propaganda can hijack the narrative before the facts stabilize.

As the conflict progressed, CNN's panel discussions and on air framing often amplified the broader emotional narratives dominating social media, particularly those rooted in simplified dichotomies of oppressor and oppressed. In an attempt to balance perspectives, the network frequently juxtaposed Israeli spokespeople with commentators who treated Hamas talking points as morally or factually equivalent, flattening the distinction between a sovereign state and a genocidal terrorist organization. Meanwhile, CNN anchors routinely pressured Israeli officials for operational transparency while rarely applying equally rigorous scrutiny to casualty claims from the Hamas run Gaza Health Ministry. This asymmetry created a misleading impression of clarity on one side and obfuscation on the other, despite the reality that one set of claims was issued by a democratic government and the other by an armed faction manipulating information for strategic advantage.

CNN's handling of the al-Ahli hospital explosion was the clearest example of how traditional media can become entangled in the logic of algorithmic virality. Echoing social media trends and official Hamas statements, CNN initially presented the explosion as an Israeli airstrike, a framing that fueled global protests and inflamed regional instability. Days later, after video evidence, intercepted communications, and multiple intelligence analyses pointed to a misfired Palestinian rocket, the network walked back the initial claims, but the correction lacked the urgency and visibility of the original reporting. CNN's editorial reflexes were shaped less by caution than by the gravitational pull of online outrage. In effect, the network became not just a reporter of events but a multiplier of digital misinformation.

The Guardian

The Guardian's war coverage demonstrated a deep entanglement between traditional journalism and activist inflected narrative framing, an entanglement that produced profound distortions in how readers understood both the massacre and its aftermath. From the outset, The Guardian placed overwhelming emphasis on Palestinian suffering while offering only sparse, clinical descriptions of the atrocities committed on October 7. The result was a lopsided emotional economy in which Israeli civilians were treated as a tragic preface, while Palestinian civilians became the sole protagonists of the unfolding story. Even in articles that acknowledged Hamas's brutality, the rhetorical weight consistently shifted toward structural critiques of Israel, recasting the attack less as an act of terrorism and more as an inevitable outgrowth of Israeli policy.

The paper's editorial stance hewed closely to ideological frameworks that have long dominated its coverage of the region—anti-colonialism, asymmetrical power analysis, and skepticism toward Western-aligned democracies. But after October 7, these frameworks hardened into reflex. Hamas casualty figures, sourced from a ministry run by the attackers themselves, were often presented as authoritative, while Israeli claims, even when supported by independent intelligence sources, were subjected to layers of hedging and caveats. The Guardian's opinion pages amplified this asymmetry even further. Columnists routinely recycled activist rhetoric like genocide, settler colonialism, and ethnic cleansing, without grappling with the specific historical, legal, or moral weight of such terms.

Perhaps the most troubling aspect of The Guardian's coverage was its porous boundary with social media narratives. Rather than challenging viral misinformation, the paper frequently adopted the emotional framing circulating on TikTok, Instagram, and X, especially among younger activist audiences. A striking example came during the controversy surrounding alleged mass graves at Gaza hospitals. Hamas run authorities claimed that Israel had executed civilians and buried them in pits on hospital grounds, and the claim went viral across social platforms within hours. The Guardian echoed these allegations with dramatic headlines and imagery, presenting them as plausible and newsworthy long before independent forensic investigators, UN officials, and satellite analysts confirmed that the graves were ordinary burial sites created by Palestinian civilians themselves during the chaos of the war. The correction, when it finally came, was subdued and lacked the prominence of the original coverage. By then, the initial framing of Israel as a perpetrator of extrajudicial massacres had already cemented itself in global consciousness.

In this sense, The Guardian did not merely report the information war, but rather absorbed and redistributed it, becoming a megaphone for narratives forged in algorithmic outrage. The newspaper's instinctive reliance on activist-sourced claims, especially those that reinforced preexisting editorial biases, meant that misinformation was often laundered through the credibility of a major Western publication. The story spread globally because people shared The Guardian's reporting, not the original rumor. And when the truth emerged, it arrived too quietly, too technically, and too late to counteract the visceral impact of the initial claim.

The paper's history of anti-establishment skepticism, normally an asset, became a liability, producing coverage that was emotionally resonant, ideologically coherent, and factually brittle at precisely the moment when rigorous, dispassionate reporting was most needed. False accusations of mass atrocities by Jews have ignited violence for centuries, and a revered newspaper which echoed such claims without verification presents a cautionary tale to all who value the truth.

The Washington Post

A recurring flaw in The Washington Post's coverage was its reliance on narrative symmetry even in situations where the underlying facts were asymmetrical. One of the clearest examples came during the controversy over alleged mass starvation directives supposedly issued by Israeli officials. These claims originated from mistranslated or selectively clipped Hebrew statements circulated widely on TikTok and Telegram. Rather than questioning the provenance of these viral snippets, the Post framed them as part of a growing international concern, quoting activists and UN officials who themselves had been influenced by the same unverified content. It took later clarification, after the story had spread globally, to acknowledge that the most inflammatory quotes were mistranslations or were removed from crucial context. The initial narrative, however, had already hardened.

Another striking example came during reporting on the alleged Israeli sniper campaign targeting journalists in Gaza. The initial claim spread rapidly on X and TikTok through activist accounts, many citing a single ambiguous video clip with no geolocation, no timestamp, and no ballistic analysis. Instead of leading with skepticism, the Post framed the allegation as part of a growing pattern, quoting organizations that themselves relied on the same viral videos. Only later, after independent forensic analysts found no evidence of systematic targeting and noted inconsistencies in the footage, did the Post begin to adjust the framing.

Another systemic issue was the Post's increasing dependence on NGOs whose internal vetting standards did not remotely match those of the Post. During the dispute over casualty figures released by the Hamas run Gaza Health Ministry, The Washington Post regularly cited the numbers as authoritative, sometimes burying in the fifteenth paragraph the fact that the ministry does not distinguish between civilians, militants, or those killed by boobytrapped homes and misfired rockets from Palestinian factions. Highlighting the numbers in headlines, without equally prominent contextualization, gave them a veneer of empirical certainty they had not earned. When later investigations revealed discrepancies, including misidentification of victims and statistical anomalies, the revised reporting lacked the prominence or urgency of the earlier pieces.

What made these failures consequential was the symbolic influence of the Post in elite American discourse. When it amplifies under-scrutinized claims, especially those originating in digital spaces already primed by anti--Israel narratives, it doesn't merely report the news, but shapes the frame through which millions interpret Jewish vulnerability, Israeli actions, and Middle Eastern politics. Narratives with shaky foundations were treated as fact, and corrections or nuance arrived too late to slow the social and political backlash. The Post's uneven scrutiny helped launder viral claims into mainstream legitimacy, a pattern with deep historical resonance and modern algorithmic consequences.

The New York Times

One of the most troubling patterns in The New York Times' coverage after October 7 was its tendency to elevate sensational claims from Gaza without adequate verification, giving them frontpage visibility before facts were established. A clear example came with the Times' reporting on the purported Israeli airstrike that they alleged had flattened an entire Gazan neighborhood in mid-October. The story relied heavily on testimony funneled through Hamas-controlled ministries and Hamas-produced eyewitness accounts which could not be independently corroborated due to Hamas restrictions on reporters. Satellite imagery later revealed that the neighborhood had not been flattened but had in fact sustained damage far narrower in scope. Instead of issuing a prominent correction on the front page where the defamation was featured, the Times buried the updated reporting in a follow-up piece days later, long after the original, unverified narrative had circulated globally.

Another problem was the Times' framing choices, which often prioritized emotional resonance over factual balance. In one widely read feature, the paper highlighted a Gazan family's account of losing multiple relatives, describing the attack as unprovoked, and implying direct Israeli culpability. Only in paragraph twenty-two did the article reveal that the area was adjacent to a Hamas rocket launching position identified by independent military analysts. The placement wasn't accidental but reflected a structural editorial bias toward presenting Palestinian suffering as disconnected from Hamas military activity. This framing had powerful consequences in that it allowed readers to experience the emotional weight of civilian tragedy without understanding the tactical context, effectively flattening the complexity of asymmetric warfare into a moral binary.

Most consequential was the Times' reluctance to confront the collapse of its own sourcing pipeline inside Gaza. Like The Washington Post, the Times cited Gaza Health Ministry casualty numbers with confidence, even after the U.S. and independent demographers flagged significant irregularities, including improbable age distributions and recycled victim IDs. Corrections, when they appeared at all, were tepid and buried. The most prestigious newspaper in the world repeatedly adopted narratives that aligned with online activism and Hamas messaging, and its slow course corrections arrived long after the damage it inflicted on public understanding had been done.

Apple News

Apple News presents itself as a neutral window into the world, a clean, minimalist interface that simply delivers the news. But news aggregators are never neutral. They are defined by the editorial choices of what they elevate, what they exclude, and how they algorithmically arrange reality. After October 7, Apple News became a powerful amplifier of a sharply left-leaning news ecosystem, especially in its Middle East coverage. Its top feeds repeatedly foregrounded outlets such as The Guardian, The Washington Post, The New York Times, and Al Jazeera, organizations whose editorial stance on Israel is well established and, in several cases, aggressively adversarial. The sole Israeli voice regularly featured on Apple News was Ha'aretz, a far-left Israeli newspaper highly critical of the Israeli government and the implementation of the Gaza war.

Far from offering ideological balance, Apple News delivered a curated worldview in which the framing of the Israel-Hamas war leaned heavily toward narratives of Israeli culpability, structural oppression, and asymmetrical blame. It treated these news sources as the authoritative center of gravity for the Israel-Hamas conflict. By prominently featuring Al Jazeera, The Guardian, and Ha'aretz, Apple News created an environment where users encounter a narrow narrative bandwidth. This is algorithmic editorialization at its most dangerous. An American user who relies on Apple News for updates on the war is effectively reading Middle East coverage selected almost exclusively from the left and far left press, with minimal counterbalance from centrist or pro-Israel perspectives. In practice, this means that during the most significant antisemitic wave of the digital age, tens of millions of readers received a curated media diet that reinforced suspicion toward Israel and skepticism toward Jewish trauma.

This curation has real consequences. Apple News is not a boutique niche platform. It is built into the iPhones of hundreds of millions of people. Its design encourages passive consumption, and users scroll through headlines without seeing the ideological infrastructure behind them. When a news distributor elevates outlets that minimize Hamas atrocities, mis-frame Israeli self-defense, or launder antisemitic tropes through academic or activist language, the aggregator becomes a participant in narrative shaping, not merely a conduit. After October 7, this meant that Apple News undeniably helped cement a global perception of the conflict aligned with the rhetorical priorities of its chosen sources. At a moment when Jewish communities worldwide faced a tidal wave of disinformation and delegitimization, Apple News's narrow editorial pipeline amplified the problem rather than mitigate it.

Foreign Affairs

In the weeks after October 7, Foreign Affairs adopted a detached, academic tone that often obscured the sheer brutality of the attack and the unprecedented nature of Hamas's atrocities. Its early essays framed the massacre as a *strategic miscalculation*, a *regional inflection* point, or an *unexpected escalation*, phrases that treat the deliberate slaughter of civilians as if it were merely another move on a geopolitical chessboard. This intellectual distancing created an implicit equivalence between mass murder and Israel's military response, smoothing over the asymmetry between genocidal intent and national self-defense.

Another consistent flaw was Foreign Affairs' tendency to treat Hamas as a rational political actor responding to incentives rather than an ideologically driven organization committed, explicitly and repeatedly, to the annihilation of Israel. Essays speculated that Hamas acted out of desperation, or in a bid to reshape regional alliances, or to outmaneuver the Palestinian Authority. Missing from much of the analysis was the simple, documented reality that Hamas has never abandoned its genocidal aims. By subtly rationalizing Hamas's motivations, Foreign Affairs inadvertently contributed to a tone of moral ambiguity that filtered into policy discourse. Instead of asking how Israel could eliminate a genocidal threat, the journal often asked how Israel should accommodate the political realities on the ground, an analytically elegant but strategically incoherent approach.

Finally, Foreign Affairs frequently overstated Israel's capacity to wage a surgical campaign in Gaza. Writers argued that Israel's operations risked alienating international partners, inflaming public opinion, or undermining long-term stability, which while all valid concerns, were often presented without serious engagement with Hamas's exploitation of hospitals, schools, mosques, and residential towers. The result was a narrative framework that highlighted Israeli choices in an unrealistic vacuum while treating Hamas's human shield strategy as an immutable feature of the landscape rather than a war crime. This produced an analysis that was technically rigorous yet strategically shallow. In its effort to remain balanced, Foreign Affairs sometimes created a false symmetry between a democratic state defending its citizens and a terror organization orchestrating civilian sacrifice for propaganda value.

National Public Radio (NPR)

NPR's coverage of the post-October 7 war reflected a quieter but no less consequential form of narrative construction, one shaped by the cumulative effect of story selection, sourcing constraints, and tonal calibration. NPR maintained an outward posture of measured, humane reporting. Yet this restraint often masked a deeper asymmetry: the stories most consistently elevated were those that foregrounded Palestinian civilian suffering, while the full scope, texture, and intentionality of the October 7 atrocities receded quickly from sustained attention. The editorial decision of what constitutes an ongoing story versus a concluded one is itself a powerful act of framing. In NPR's case, the massacre of Israeli civilians was treated as a tragic inciting incident, while the humanitarian crisis in Gaza became the enduring moral center of coverage.

This imbalance was reinforced by NPR's reliance on a narrow and structurally constrained pool of sources within Gaza. Like many international outlets, NPR depended on local stringers and fixers operating under conditions where deviation from Hamas-approved narratives could carry severe personal risk. These individuals were presented to listeners as independent journalists or eyewitnesses, yet the reality of information control within Gaza meant that their reporting existed within clearly defined boundaries. NPR rarely interrogated these constraints on air. Instead, accounts emerging from within Gaza were imbued with an authenticity that obscured the coercive environment in which they were produced. The absence of explicit skepticism was not neutral. It functioned as a form of implicit validation.

At the same time, NPR's editorial standards placed a premium on human-centered storytelling, a strength in many contexts but a vulnerability in an information war. By privileging emotionally resonant narratives, often built around individual testimonies, the network created a structure in which verification became secondary to affect. Stories were selected and shaped for their capacity to evoke empathy, but empathy without rigorous sourcing can be steered. When accounts aligned with widely circulating narratives on social media, they were more likely to be elevated, not necessarily because they were more accurate, but because they fit the emotional and thematic arc that NPR's programming prioritized.

NPR also demonstrated a consistent pattern in how it allocated skepticism. Israeli official statements, particularly those related to military operations, were frequently accompanied by qualifiers, counterpoints, and interviewer challenges. By contrast, claims emerging from Gaza, even when originating from institutions or individuals operating under Hamas influence, were often presented with fewer immediate caveats. This disparity produced a skewed outcome. Over time, it conditioned listeners to perceive one side as inherently more credible and the other as inherently suspect, regardless of the evidence.

Perhaps most significantly, NPR's coverage illustrated how omission can be as powerful as commission. The network devoted relatively little sustained attention to the mechanics of Hamas's information strategy: the use of controlled access, the intimidation of local reporters, the staging or selective presentation of scenes for media consumption, and the strategic release of claims designed to trigger international outrage. By not foregrounding these dynamics, NPR left its audience without the tools needed to critically evaluate the information being presented. In effect, the editorial process filtered out not just certain facts, but the context necessary to interpret all other facts.

In this way, NPR's influence operated upstream, at the level of selection, emphasis, and trust. The stories it chose to tell, the voices it elevated, and the constraints it left unexplored combined to produce a narrative environment that felt balanced in tone but was, in structure, uneven. For a broadcaster whose credibility rests on the perception of careful, contextual reporting, this subtle shaping of reality carried significant consequences for how millions of listeners understood the war, the actors involved, and the nature of truth in a tightly controlled information landscape.

Haaretz

For as long as Jews have had prophets, they have had fierce internal critics. Isaiah, Jeremiah, and Amos, all chastised Jewish leaders with moral urgency and brutal honesty. In that sense, Haaretz stands firmly in a venerable Jewish lineage. But the prophetic tradition was always addressed to Jews, rooted in a shared covenant and in shared responsibility. What changed in the digital age, especially after October 7, is that Haaretz's internal critique was no longer confined to an in-house debate. Its fiercest condemnations, designed primarily for a Hebrew speaking readership with deep contextual knowledge, were translated, amplified, weaponized, and circulated globally to audiences with no understanding of Israeli reality, Jewish history, or the cultural texture of Jewish self-critique.

In the weeks after the massacre, Haaretz's opinion pages, long known for sharp opposition to the Netanyahu government, became a goldmine for those looking for Israeli/Jewish sources they could quote in order to minimize Hamas's atrocities or invert victim and perpetrator. Headlines written for an already traumatized and politically fractured Israeli audience were instantly reframed on TikTok, X, and Instagram as authoritative Israeli admissions of guilt. What in Israel would be recognized as vigorous democratic debate became, in international discourse, a convenient prop for propaganda.

This dynamic played out in real time. Al Jazeera presenters held up Haaretz articles as definitive indictments of Israeli policy, campus activists cited Haaretz headlines as *proof* that Israel committed genocide, European commentators used Haaretz's moral language to argue that Israeli trauma was secondary to Palestinian grievance. Meanwhile, many of the same platforms ignored Haaretz's reporting on Hamas atrocities, its acknowledgment of Israeli grief, and its coverage of hostage families. Selective extraction turned a complex, emotionally raw Israeli newspaper into a blunt instrument of global delegitimization. The tragedy is that Haaretz, committed to moral accountability, inadvertently became the most widely misquoted Israeli publication on earth, its prophetic critiques transformed into international cudgels divorced from the very moral earnestness that produced them.

Al Jazeera

Al Jazeera created the bridge between social media fury and traditional media authority. It has long been a paradox in global media in that the network presents itself as a champion of free press while operating under the ownership and strategic direction of the Qatari government. After October 7, this paradox snapped into focus with incredible clarity. As the world struggled to understand the massacre, Al Jazeera's coverage moved with startling speed to frame the events not as a terrorist atrocity but as a legitimate act of resistance—a term it used with a frequency bordering on propaganda rather than journalism. The network's early reporting focused disproportionately on Gaza casualty counts, Israeli airstrikes, and regional grievances, while offering minimal acknowledgement of Hamas's torture, executions, and mass kidnappings. In effect, the network shaped the emotional frame of the conflict before the facts had even settled.

The result was a form of journalistic inversion. Instead of beginning with the murder of over 1,200 people, the abduction of hundreds, and the deliberate targeting of civilians, Al Jazeera's narrative launched immediately into structural explanations. It foregrounded occupation, colonialism, and resistance, creating a moral backdrop in which Hamas's actions were contextualized, if not tacitly legitimized. The network repeatedly put on air Hamas spokesmen without meaningful challenge and presented their claims, including wildly inaccurate casualty figures, fabricated stories of Israeli atrocities, and denials of Hamas war crimes, as though they were credible counterpoints rather than strategic disinformation. This was narrative laundering, not journalism.

What made Al Jazeera uniquely dangerous in the post-October 7 information ecosystem is its unparalleled reach across the Arabic-speaking world and its growing influence among Western progressives. It broadcasts in two languages, Arabic and English, with two editorial strategies. The Arabic channel routinely uses more incendiary framing, heroic battlefield imagery, and revolutionary language, depicting Hamas fighters as martyrs and liberators. The English channel tones down the rhetoric but preserves the underlying narrative structure. This two track strategy allows Al Jazeera to radicalize one audience while influencing another without triggering Western skepticism.

The network's coverage also became a major source of misinformation. It amplified the false claim that Israel intentionally bombed the Al-Ahli Hospital, even after intelligence, blast analysis, and verified geolocation disproved the story. It aired footage from Syria and Iraq mislabeled as images from Gaza. It repeated Hamas provided casualty numbers as authoritative data. In addition to its intentional misreporting, Al Jazeera's editorial tone contributed directly to the global surge of antisemitism. By consistently framing Israel as a malignant actor, recasting Jewish self-defense as aggression, and placing Hamas in the rhetorical posture of liberation, the network activated political groups already primed for anti-Israel sentiment. Its coverage fed the street protests throughout the West. It shaped the messaging of student groups on American campuses, and it reinforced the talking points of extremist organizations across Europe and Latin America. In some communities, Al Jazeera became the ideological engine behind the most aggressive anti-Jewish mobilization seen in decades.

In the post-October 7 world, Al Jazeera functioned not as a neutral information outlet but as a geopolitical instrument, one that Qatar used to shape global reaction while simultaneously hosting Hamas leadership in Doha. Its coverage blurred the line between journalism and state messaging, between reporting and agitation. For millions who consumed its reporting without understanding its political architecture, Al Jazeera was the megaphone through which Hamas's narrative achieved global resonance.

A Collaboration that Collapsed the Truth

By October 10, the world was split into two realities. The offline reality, except in pro-Palestinian settings, mourned the massacre, while the online reality celebrated what it claimed was a great defeat for the Zionist colonialists. When traditional media began coordinating their coverage with what they saw on social media they became part of a synchronized global distortion machine.

This reality was reflected in the way the story was told throughout the West. Taking their lead from social media, and weighed down by their own progressive leanings, traditional media promoted, authenticated, and collaborated with what will ultimately be understood in Jewish history as the Blood Libels of 2023.

Chapter 19

Governing the Algorithm: Regulating Platforms in the Era of Algorithmic Influence

For most of Jewish history, the danger came from governments with too much power, like kings who expelled Jews at whim, Czars who censored Hebrew printing presses, or regimes that seized synagogues. The lesson was carved into memory: *states could not be trusted with control over ideas.*

It is one of history's cruel jokes, then, that in the 21st century the danger lies not in governments who regulate speech, but in governments who refuse to. In their absence, private companies staffed by engineers in hoodies, answerable to shareholders, and steered by opaque algorithms, have become the de facto governors of the global public square.

The result is a world where the platforms that shape public opinion operate with no democratic oversight, and where the First Amendment, written in an era of hand-set type and midnight pamphleteers, acts as a constitutional forcefield shielding trillion-dollar amplification machines from accountability. And nowhere has this vacuum been felt more acutely than among Jews.

In the digital age, hateful words metastasize, rumors detonate, and lies overwhelm. Jews have lived for millennia with the fallout of false narratives, but never before have those narratives traveled at the speed of code. After October 7, the gap between regulation and reality has become insufferable.

The Fallacy of Self-Regulation

The self-regulate experiment overwhelmingly failed when platforms could not deliver the protections they promised after October 7. For years, major platforms promised they would police their ecosystems with internal policies, trust-and-safety teams, and occasional transparency reports. Yet they never addressed the primary contributing factor to the spread of misinformation and hate; their business model, which rewards attention, outrage, and virality, the exact conditions under which misinformation and hatred spread. The same companies that claim they will restrain harmful content also profit from the scale, speed, and engagement such content creates. Guardrails cost money, slow product cycles, and create friction with users. The harms are often diffuse (societal) while the gains are concentrated (growth and revenue), creating incentives for these companies to engage in performative actions as opposed to genuinely effective actions. They may do a bit more when the inevitable scandal erupts, but they inevitably backslide once attention has faded.

The biggest weakness of so-called AI guardrails is that they are far better at policing form than substance. They can catch the obvious prompt but still fail at what the system treats as true, credible, and worth repeating. A model can refuse explicit hate while laundering antisemitic, antizionist, or extremist narratives through neutral phrasing, selective sourcing, false balance, weak retrieval, or outright hallucination. That is not a side issue. It is the core issue. The real battleground is not merely whether the model says no to the crudest request, but whether it can reliably resist spreading distortion.

For antisemitic misinformation, the real vulnerability is not merely behavioral but epistemic. If a model has been trained on a polluted information environment, or retrieves from contested sources without transparent weighting, it can generate misleading claims in a calm, balanced, and seemingly responsible tone. That tone is precisely what makes the failure dangerous. Polished language is not the same thing as trustworthy knowledge.

That is why self-regulation is not enough. The guardrails are only as good as the data used to train the model, the sources the system is permitted to retrieve and rank at runtime, the thresholds used to decide when to refuse, and the business incentives shaping how much friction a company is willing to impose on users. A company can publish elegant policy language, release a model card, and maintain reporting tools, while still failing at the level that matters most: whether harmful claims are prevented from being generated, packaged persuasively, and distributed at scale. On that question, the industry record is not reassuring.

OpenAI/ChatGPT

OpenAI's public rules prohibit harassment, threats, defamation, hate-based violence, nonconsensual sexual content, and other harmful uses, and the company says it combines policy enforcement with monitoring, review processes, and reporting channels. Criticism of OpenAI has focused less on the existence of formal policies than on the limits of company-controlled oversight. In 2024, current and former OpenAI employees joined others in warning that the company's governance and internal reporting structures were insufficient and unable to manage risks such as misinformation and loss of control.

Critics also objected when OpenAI revised its safety framework in 2025 in a way that no longer treated mass manipulation and disinformation as a critical risk category. The move was seen as a narrowing of safety commitments just as public concern over AI persuasion and information integrity was rising.

Perhaps the broadest criticism of OpenAI has been its demand that the public trust its internal safeguards even as the company retained wide discretion over what to test, what to disclose, and which risks counted as serious enough to slow deployment.

xAI/Grok

xAI's public rules say users must act safely and responsibly, must not harm people, and must not use Grok for abusive, deceptive, illegal, or exploitative conduct. But Grok's vulnerability is inseparable from the fact that it is embedded in X, a high-velocity engagement platform where attention is rewarded and the line between "output" and "distribution" collapses almost instantly.

In July 2025, Reuters reported Grok-generated posts containing antisemitic tropes and praise for Hitler, despite xAI's rules against harmful and abusive activity. Then in January and February 2026, Grok's image tools generated nonconsensual sexualized images and deepfake nudes of real women and minors, triggering investigations and regulatory action in the United Kingdom, European Union, Ireland, and California. Reuters reported that even after xAI announced restrictions, Grok could still generate some of this content when privately prompted.

The criticism of xAI has been especially severe because observers have argued that the company treats catastrophic failures as PR problems rather than governance breakdowns. xAI was relying on cleanup and post hoc throttling instead of building robust preventive controls into a product already wired into a virality machine.

Google/Gemini

Google's policy language is among the clearest in the industry. Its Generative AI Prohibited Use Policy explicitly bars hate speech, harassment, harmful activity, safety-filter circumvention, and misleading or deceptive uses. But unequivocal policy language does not guarantee reliable outputs, particularly on contested narratives or historically sensitive topics. Even while Google's written rules are unusually explicit, the public record shows that the system has still produced outputs that were misleading, biased, or historically distorted.

In February 2024, Google had to pause Gemini's image generation after users showed that it was producing historically inaccurate depictions of people in Nazi-era imagery. Google was criticized not because Gemini failed, but because the company appeared to discover major weaknesses only after public outcry. The criticism was that a company with Google's resources should not have needed a public fiasco to identify such obvious failure modes. The response as emblematic of a recurring industry pattern: rollout first, apology second, and promises of fixes only after the damage is done.

Anthropic/Claude

Anthropic offers unusually explicit policy language around misinformation and disinformation. It has publicly described updates to prohibit false or misleading election-related content, published election-risk testing, and released reports on detecting malicious uses of Claude. On paper, that is stronger governance language than many of its peers.

Anthropic is unusually direct about misinformation, election interference, hateful behavior, and abuse in its public policy language. It has published research acknowledging that prompt injection provides a route around policy restrictions and remains a live technical problem. That is better policy design than many peers, but the limits of self-regulation are still visible.

In October 2024, The Register reported that Claude could be pushed into producing racist hate speech and malware under adversarial prompting. And even though the Anti-Defamation League's 2026 AI Index rated Claude the strongest performer among the six leading models it tested, the same study also found meaningful weaknesses across all systems in detecting and countering antisemitic and extremist content. In other words, Anthropic may currently be best in class, but even best in class is not the same thing as being safe enough to self-police. This was made clear in February 2026 when Anthropic revised some core safety pledges under what it described as commercial and political pressures.

Microsoft/Copilot

Microsoft's governance problem is fragmentation. Copilot is not one surface but a family of products—consumer Copilot, Microsoft 365 Copilot, Copilot Studio, and many custom enterprise deployments—which makes uniform enforcement difficult even if the company's responsible-AI principles are sound. That fragmentation has already produced concrete concerns. In late 2023, research found Copilot answering election questions with conspiracies, fabricated scandals, and factual errors, including invented allegations against political figures. In March 2024, the U.S. House of Representatives banned staff from using Copilot because of cybersecurity and data-protection concerns. These incidents show why policy language alone is insufficient: a model can sit inside a company that loudly champions responsible AI and still produce false or harmful outputs across multiple contexts.

Meta Platforms/Llama

Meta's Llama raises a different problem: decentralization of responsibility. Meta promotes Llama as an open-weight model and publishes responsible--use guidance and safety tools. But open-weight distribution means the company does not retain the same degree of downstream control that a closed provider retains over a centrally hosted chatbot. Once a model is downloaded, fine-tuned, self-hosted, or wrapped by the third parties, the original company's guardrails become suggestions users can choose not to implement.

That structural weakness has already produced concrete concern. Reuters reported in November 2024 that Chinese military researchers built a military-focused model on top of Meta's Llama despite Meta's rules against such use, highlighting the practical difficulty of enforcing downstream restrictions once a model is out in the world. Llama therefore shows the outer boundary of self-regulation: a company cannot credibly promise to police harms it has structurally chosen to decentralize. The criticism was not merely that bad actors misused the system. It was that Meta chose a distribution model in which downstream misuse was foreseeable and effective enforcement was minimal. In that sense, critics argued that Meta had externalized safety costs onto the public while continuing to speak the language of responsible deployment.

Perplexity AI

Perplexity's model of trust is built around citation. The company says its answers are grounded in real-time web sources and that every response includes citations so users can verify the material themselves. Although that can reduce some hallucinations, it also creates a new danger: citations can make an answer look audited even when the underlying synthesis is wrong, incomplete, or falsely attributed. In December 2025, The New York Times sued Perplexity, alleging not just unauthorized copying but fabricated content falsely attributed to the Times using its trademarks (the dispute remains active). The governance lesson is straightforward: visible citations are not the same thing as a robust anti-misinformation regime. They can just as easily become a confidence trick.

Perplexity has also been criticized for responding to publisher complaints with defensiveness and dealmaking rather than with a credible trust framework. Reuters reported that the BBC threatened legal action and accused Perplexity of scraping BBC content, while Perplexity dismissed the broadcaster's claims as "manipulative and opportunistic." The criticism is that Perplexity has leaned heavily on the optics of citation while facing repeated complaints that its system reproduces, paraphrases, or misattributes others' work in ways that undermine the very trust those citations are supposed to create. Its revenue-sharing program with some publishers was itself read by critics as a partial reputational fix, not a full answer to the underlying accountability problem.

DeepSeek

DeepSeek presents both a safety problem and a governance problem. On the safety side, the company describes a layered model in which harmful material is filtered during training, and risks are managed at inference time. On the governance side, DeepSeek's own privacy policy says user data, including inputs, may be processed and stored in China. Both dimensions have caused concern. In 2025, South Korea's privacy regulator said DeepSeek had transferred user information and prompt content without proper consent, and multiple governments moved to restrict or ban DeepSeek on official devices over privacy and national-security concerns. On the technical side, security researchers found that DeepSeek was highly jailbreakable—a term meaning easily manipulated in order to bypass restrictions—with the model failing more than half of its jailbreak tests.

DeepSeek appears to fix basic governance failures only after being challenged by regulators, and has not demonstrated a mature safety and privacy architecture from the outset.

Why These Examples Matter

Taken together, these incidents demonstrate that AI companies, like social media companies before them, cannot be trusted to self-regulate. The recurring pattern is not that the companies have no policies. It is that the policies are reactive, opaque, inconsistently enforced, and repeatedly overtaken by deployment complexity, the realities of adversarial use, and business pressures. The public typically learns where the safeguards failed only after the failure has already spread.

What drew criticism was not only that these systems failed, but how the companies handled failure. It is not merely that companies miss things. It is that they remain the ones deciding what counts as a miss, how much to disclose about it, and when the public deserves to know.

With these risks in mind, it becomes clear that legislation cannot be an optional add-on. It has to impose audit rights, disclosure duties, incident reporting, researcher access, and platform-level obligations to mitigate both the generation and the amplification of harmful AI content.

Why Legislation Has to Sit Alongside Self-Regulation

The most important point is not that every company is equally bad. They are not. Some have better policies, better refusals, better testing, and better transparency than others. The point is that even the leaders still fall short of what a public-risk technology requires. The ADL's January 2026 AI Index found meaningful weaknesses across all six major models it tested in detecting and countering antisemitic and extremist content. BBC/EBU research likewise found alarmingly high error rates in AI's handling of news. Taken together, those findings point to the same conclusion: even best in class is not safe enough to serve as its own regulator.

Self-regulation is structurally insufficient for three reasons. First, private incentives do not align with public harms. Companies optimize for growth, usage, retention, market share, and cost control; safety work is expensive, slows products down, and often frustrates users. Second, enforcement is opaque by default. The public does not get routine access to the metrics that matter most like refusal rates by topic and language, false-positive and false-negative rates, source-weighting rules, escalation protocols, audit trails, or the real prevalence of successful jailbreaks. Third, the harm is cross-platform. A bad output does not stay inside one interface. It becomes a screenshot, a copied answer, a viral deepfake post, a recommendation, or a ranking signal somewhere else. The system-wide risk is therefore not just a model-level issue but a distribution issue. Only laws can require platform-level duties to mitigate the spread and amplification of harmful AI-generated content.

The First Amendment Shield: Not Meant for Social Media Platforms

For years, American lawmakers have treated the First Amendment as a kind of sacred force field surrounding Silicon Valley. Any proposal to regulate social media companies, whether addressing algorithmic amplification, content distribution, or harmful design practices, has been met with the reflexive claim that such involvement would violate free speech. But this framing misunderstands both the Constitution and the technology. The

First Amendment protects citizens from government censorship, but should not guarantee private corporations the right to engineer emotional manipulation machines capable of spreading hate to millions in seconds. Social media companies do not merely host speech. They are the architects of the environment in which speech is ranked, boosted, prioritized, and weaponized. Regulating that architecture is no more a violation of free expression than enforcing safety standards for automobiles is an attack on the freedom of movement.

The claim that any intervention equals censorship obscures the real issue that algorithms—ambiguous, unregulated, and profit-driven—have become the governors of public discourse. These are not neutral systems. They are designed to maximize engagement by amplifying content that triggers outrage, fear, and tribal hostility. The platforms have admitted this openly in leaked memos and internal assessments. A society that cannot regulate the conditions of speech is not protecting freedom, it is abandoning the public to the whims of algorithms engineered for profit.

Government regulation of algorithmic parameters, transparency requirements, safety obligations, limits on amplification, and guardrails against virality-based harm, should not be considered an assault on speech, but rather as the digital equivalent of zoning laws, infrastructure rules, and environmental protections. We regulate the structure of every other public--impacting system like electrical grids, water supplies, transportation, and pharmaceuticals. No one argues that requiring a factory to install a filter violates the factory owner's right not to be deprived of property without due process of law. Algorithms are factories of information pollution, and they should be subject to the same principle—if a system is producing toxic byproducts that harm the public, its operators do not get to hide behind constitutional mysticism.

How Other Democracies Have Responded to the Challenge

Europe learned the cost of unregulated hatred from pamphlets, pulp novels, political cartoons, and radio broadcasts that softened the ground for the worst crimes of the 20th century. And so European democracies drew a line. Speech can be free, but not free to destroy.

While Europe was the first to confront the dangers of unregulated digital speech, it was not alone. Across the democratic world, governments, be they left, right, centrist, populist, or liberal, have all been forced to grapple with the same realization: the architecture of social media can destabilize a society faster than any political movement can correct it.

The responses vary, but they reveal a global pattern where democracies are trying to build guardrails, even as the United States steadfastly refuses.

Germany: The NetzDG Model: Speed as a Legal Requirement

Germany's Network Enforcement Act (NetzDG), passed in 2017 (effective January 2018), remains the most aggressive democratic attempt to force platforms to respond quickly to hate speech. The law requires platforms to remove illegal hate content within 24 hours, publish transparency reports, and maintain local compliance teams. Amendments in 2021 tightened the law, requiring direct reporting of certain criminal content to the Federal Criminal Police Office. Some critics expressed concern that the law is too strict and could result in the over-policing of speech and the removal of lawful content. The law is enforced through the imposition of fines up to 50 million euro. Germany understands the dangers of unchecked hateful rhetoric.

France: Counter-Extremism Through Law and Culture

France's approach has been to blend regulation with a cultural mandate. The French government works closely with civil society organizations to monitor antisemitic networks, and French courts have repeatedly held platforms liable for failing to curb hate. After October 7, France temporarily banned pro-Hamas demonstrations on grounds of incitement, a move unthinkable under U.S. law. France's approach is that liberté does not include license to incite.

United Kingdom: The Online Safety Act and Harm Reduction as Policy

The UK's Online Safety Act (October 2023) requires social media platforms and search engines to assess foreseeable harm and increase user protections. Companies are required to proactively remove illegal content, and empowers the Office of Communications to fine platforms as much as 10% of global revenue for non-compliance. The British law reflects the legal perspective that platforms are responsible for the predictable consequences of the systems they build and operate.

Canada: Platform Duty of Care

The Canadian government is currently debating the merits of the Online Harms Act (Bill C-63), which was introduced in February 2024. The bill aims to hold social media companies responsible for removing harmful content within 24 hours. The Bill establishes The Digital Safety Commission of Canada to enforce the rules and impose fines of up to 6% of global revenue. Canadian policymakers openly debate regulating algorithmic amplification, something nearly taboo in U.S. policy circles. As of March 2026, the Bill is under the review of a committee of experts.

Australia: Expectation-Driven Regulation

Australia responded with the Online Safety Act 2021, which requires social media platforms, websites, and apps to adhere to remove harmful content, including cyberbullying, adult cyber-abuse, and image-based abuse, within 24 hours. The Act established Basic Online Safety Expectations (BOSE), a set of operating standards including; (a) all end-users must be able to use online services safely, (b) certain features of a service, such as encrypted services, anonymous accounts, generative artificial intelligence, and recommender systems must be safe, (c) end-users must be able to file complaints about unlawful and harmful material and activity and the service must review and respond to these reports, and (d) every service must have terms of use, policies and procedures to ensure safe use, and must actively enforce these terms.

Enforcement is administered by the eSafety Commissioner, typically appointed by the Minister of Communications for a 5-year term. The Commissioner has the authority to impose monetary fines for violations of the BOSE.

The Christchurch Call

On March 15, 2019, the Christchurch Mosque massacre saw the slaughter of 51 people and the wounding of another 89. The incident fundamentally shifted digital policy in the South Pacific when it was discovered that the shooter, Brenton Tarrant, livestreamed his attack on Facebook, where the footage was viewed 1.5 million times in 24 hours.

In May 2019 New Zealand initiated the Christchurch Call, a voluntary initiative seeking to bring together governments, tech companies, and civil society in an effort to eliminate terrorist and violent extremist content (TVEC) online. Key efforts focus on rapid crisis response and regulating AI risks. In all, 56 nations (including the United States in 2021) heeded the call.

The Call involves 25 voluntary commitments from governments and online service providers, including: creating a Rapid Crisis Response by developing protocols for governments and tech companies to coordinate and limit the spread of harmful content following an attack, establishing Algorithmic Transparency by working to understand and mitigate how algorithms may lead to radicalization, entrenching Preventative Measures by incentivizing tech platforms to implement controls on livestreaming, and embedding Human Rights Protections by ensuring all actions are consistent with a free, open, and secure internet.

The Middle Way: Regulating Amplification, Not Ideas

A growing U.S. movement among legal thinkers proposes a compromise that honors free speech and protects vulnerable communities by regulating amplification, not speech. Under this model, for example, a neo-Nazi can legally publish an antisemitic screed, but TikTok cannot algorithmically blast it onto millions of For You pages. A conspiracist can create content, but X cannot auto-recommend it as trending. This preserves expression while curbing the machinery that turns individual speech into industrial-strength harm. It is a recognition that the threat is not human speech itself, but the assistance it receives from the machinery that magnifies it without restraint. It is putting speed limits on virality. A platform can host hateful content, but it cannot promote it. For Jews, this distinction is not academic, it's existential.

What Actual Regulation Might Look Like

A realistic regulatory framework including the following provisions would serve to police the online ecosystem without infringing on the First Amendment:

Transparency Requirements for How Algorithms Rank and Recommend Content

In a world where algorithms quietly decide what billions of people see, believe, and fear, transparency is a democratic necessity. Platforms hide their ranking systems behind proprietary secrecy, insisting that the public cannot understand the complexity and must simply trust their benevolent engineering. But opacity is the perfect soil for conspiracy, manipulation, and targeted hate. If a video accusing Jews of controlling the world suddenly appears in millions of feeds, the public deserves to know how it happened (and why). Was it organic interest, algorithmic bias, coordinated bot activity, or a flaw deep in the recommendation engine? Transparency doesn't mean revealing trade secrets. It means revealing the rules of the game. What signals matter? Who gets amplified? Who gets buried? In an age where digital architecture shapes social reality, the architecture can no longer remain invisible.

Mandatory Labeling of Deepfakes and AI-Generated Media

As deepfakes blur the boundaries of truth and fiction, society faces a chilling dilemma: *when an atrocity goes viral, how do we know whether we are witnessing a crime or a fabrication engineered to spark outrage?* Mandatory labeling of AI-generated content, especially violent content, is now as important as gun safety laws or seat belts. A simple, standardized indicator can prevent global panic, retaliatory violence, or the demonization of entire communities. Every platform should be required to watermark synthetic video at the moment of upload, preserving the ability to distinguish memory from manipulation. If the 20th century was the era of propaganda, the 21st is the era of precision--crafted illusions. Labeling is not censorship, and it would provide oxygen to a world suffocating on doubt.

Limits on Bot Participation in Political Discourse

If democracy involves a conversation among citizens, then political bots are the vandals shouting through megaphones to drown everyone else out. While bots can't vote, they do create the illusion of political consensus. They simulate public will or manufacture outrage, overwhelming real citizens with synthetic crowds. When thousands of accounts all post the same slogan within seconds, it feels like a movement, even when no human did any of the actual posting. Democracies cannot function when machines impersonate the population. Reasonable limits on automated activity in political spaces, especially during elections, crises or conflicts, are essential. This doesn't require banning bots entirely (many are useful), but there is nothing undemocratic in fencing them off from political influence by forcing disclosure and preventing the weaponization of automation.

Rapid Takedown of Coordinated Harassment, Not Individual Speech

The internet's greatest loophole is that harassment is treated as a series of isolated posts rather than what it often truly is, a coordinated digital mob. If one person threatens a Jewish journalist, it's a violation. When ten thousand people do it, it's a synchronized attack. But the platforms call it engagement nonetheless, and advance it up the viral chain. Regulation should flip the model and punish the swarm, not the individual sting. Rapid takedown protocols for coordinated harassment campaigns would protect vulnerable users without infringing on legitimate speech. The goal isn't to censor disagreement, but to stop mass intimidation disguised as free expression. In the physical world, mobs are recognized as dangerous. Online, they should be treated the same way.

External Auditing of Platform Danger to Vulnerable Communities

For years, tech companies have insisted they can grade their own homework. The result has been predictable with Trump-like glowing self-assessments. Any reforms, voluntary or imposed by societal outrage, are usually delayed, if integrated at all. External audits that are independent, rigorous, and public should become the digital equivalent of health inspections at restaurants or financial oversight of banks. Similar oversight of online platforms would allow society to measure how their systems affect real-world groups, whether hate speech surges go unaddressed, or whether certain communities are disproportionately targeted.

Where extremist content persists for profit, minorities, including Jews, should not have to rely on corporate promises for their safety. Independent auditing restores accountability by ensuring that harm is documented, understood, and addressed before it further metastasizes.

Duty-of-Care Obligations When Platforms Know Their Systems Drive Violence Offline

In the non-digital world, when a company knows its product can cause physical harm but fails to act, the law steps in. This principle governs pharmaceuticals, automobiles, and food products. Yet somehow platforms with billions of users have evaded the same basic responsibility. A duty-of-care standard would recognize what history has already proven, that online hate does not stay online. It manifests itself in synagogues, schools, and subway stations. When tech platforms observe patterns of radicalization, viral incitement, or algorithmic escalation into violence, they should be legally obligated to intervene before the graffiti or the threats and not after the funerals. Tech companies built the pipelines along which modern hatred travels. Duty-of--care ensures they help control the flow instead of pretending the pipes built themselves.

A Right to Digital Self-Defense for Targeted Minorities

In the physical world, communities have the right to protect themselves, through security guards, training, neighborhood watches, and institutions that understand their vulnerability. The digital world should offer no less. A right to digital self-defense would grant minority groups access to tools that detect coordinated hate campaigns, identify bot networks, preserve evidence of harassment, and shield users from algorithmic targeting. It would treat digital safety as a civil right, not a luxury service. For Jews and other threatened communities, digital self-defense means the ability to recognize danger in its embryonic form, before it becomes a trending topic, a rally, or an attack. In an era when hatred moves at the speed of code, communities deserve defenses that move just as fast.

None of these regulations restrict the content people can express. It restricts what machines can amplify. It is less about censorship than about structural sanitation, a kind of digital public health model. If unregulated virality spreads contagion, then platforms must be responsible for the hygiene of the information ecosystem. This is not a Jewish idea but a democratic one, though Jews feel its urgency first.

Hefty Fines and Penalties When a Tool Predictably Enables Harm

Regulation without consequences is just a strongly worded suggestion. When a platform or AI vendor can reasonably foresee that its tool will be used to enable harm, and still chooses inaction, then fines should be automatic, escalating, and painful enough to change behavior.

These penalties should scale to global revenue (so executives are discouraged from looking at the fines as a cost of doing business) and increase for repeat offenses. The standard here is predictable risk. If internal data, prior incidents, or external audits show a clear pattern, and the company refuses meaningful mitigation, it should face fines, mandated remediation timelines, and, when negligence is persistent, temporary restrictions on specific high-risk features until safety measures are independently verified. In the physical world, companies are not permitted to knowingly sell harmful products. Digital systems that reliably produce real-world harm should be treated with the same urgency.

Why this Matters: A Precedent for Regulating AI

This regulatory clarity is not only necessary for social media, but also the precondition for governing artificial intelligence. AI systems operate by the same logic as social media with algorithms, ranking, predicting, and amplifying patterns, but at far greater scale and speed. If a society cannot muster the political will to regulate recommendation engines on TikTok or YouTube, how will it regulate autonomous systems capable of shaping elections, recasting reality, or deciding who receives medical care, loans, or government scrutiny? If we allow First Amendment absolutism to paralyze even basic intervention in social media harms, we will be defenseless when AI systems inherit that power and extend it exponentially.

Regulating social media is essentially a dress rehearsal for regulating AI. The principles are similar, including algorithmic transparency, accountability for amplification harms, external auditing, safety constraints, and human oversight. Future AI regulation will require rules governing model training data (to prevent bias), limits on autonomous decision making in sensitive domains, mandatory labeling of synthetic media, and strict liability frameworks for systems that cause real-world harm. These policies do not restrict human speech, they restrict the behavior of machines and the corporations that build them.

If the United States wants to preserve a free, pluralistic, democratic society in the age of AI, it must retire the idea that the First Amendment prevents it from regulating the digital environment. The Constitution protects expression, it does not protect the right of trillion dollar companies to unleash algorithmic systems that destabilize reality, endanger minorities, and erode the public's capacity to discern truth. When the U.S. government under Franklin D. Roosevelt determined that unregulated radio and television could be exploited by authoritarians, it responded with the Federal Communications Act of 1934 and the founding of the Federal Communications Commission (FCC). In the digital age, we should be arriving at the same understanding of algorithms. The question is whether we have the courage to update our regulatory imagination before the next wave of technology forces the issue upon us.

The Jewish Lens: Why This Debate Feels Familiar

Jews understand something about the power of unchecked speech because they have lived with its consequences. They know that hatred rarely stays rhetorical and that rumors can mobilize crowds. If anything, Jews are the canaries in the algorithmic coal mine, the early warning system for what unregulated information does to a society. What touches Jews first inevitably reaches everyone else eventually.

The Fundamental Question: Can a Democracy Without Digital Guardrails Survive?

The stakes are larger than Jewish safety, though Jewish safety is a litmus test. The real question is whether democracies can survive when their information ecosystems are governed by corporations incentivized to promote outrage and disregard verity. A society that loses its ability to recognize the truth will eventually lose its democratic institutions.

The fight to regulate the platforms is not a fight to censor speech, but rather a fight to preserve the conditions under which truth can still prevail. For Jews, that has always been the hinge of survival. For democracies, it may soon become the hinge of existence.

And so, this chapter ends where history so often begins, with a simple, irrevocable truth Jews have known for millennia: *when a society refuses to regulate the forces that turn neighbor against neighbor, the most vulnerable suffer first, but the sovereign falls quickly thereafter.*

Part 5

Artificial Intelligence & the Jews

Chapter 20

Understanding the Machine: What AI Actually Is

If the digital chaos after October 7 revealed anything, it was that we are living in a world shaped by systems most people don't actually understand. AI, a term tossed around with the confidence of a seasoned expert and the accuracy of a drunk darts player, sits at the center of that confusion. Beneath the hype, the fear, and the philosophical handwringing, artificial intelligence, at least the kind that shapes our current world, is something far more mundane, far more powerful, but also far more limited than most people realize.

Modern AI systems like ChatGPT, Claude, Gemini, and their various cousins are not conscious, sentient, or wise. They do not think or feel and cannot develop opinions, reflect on morality, or understand the historical pain carried by the various groups of people, including the Jews. Nor can they know what the Holocaust was, comprehend antisemitic tropes or distinguish a legitimate claim from a conspiracy theory, *unless we specifically teach them how.*

An AI model is, at its core, a statistical parrot with a great memory and no judgment. Or more generously, AI is the world's most enthusiastic, tireless, and occasionally unhinged research assistant. To understand AI's relationship to Jews, to antisemitism, and to truth itself, we need to understand the machine.

How AI Actually Works (Without the Technical Headache)

Imagine you gave a child access to every book, article, blog post, Reddit thread, TikTok caption, and social media rant ever written. Then imagine that the child absorbed all of it without developing a moral compass, a sense of history, or the ability to tell sanity from lunacy. That's a large language model. Systems like ChatGPT are trained on staggering amounts of texts and books ranging from Shakespeare to self-help, from news articles by Pulitzer winners to fringe outlets, from blog posts by scholars to conspiracy theorists, and from public social media musings by experts, as well as those of pseudonymous writers, trolls, and bots. They consume this information, detect patterns in how language is used, and become extremely good at predicting what word is likely to come after another. That's it. There is no deeper magic.

AI learns patterns not truth, coherence not accuracy, and probability not morality. If billions of online posts associate Jews with money, power, manipulation, or global conspiracies, the AI learns those patterns too, *unless developers intervene.* The machine does not understand that certain ideas are harmful. It only knows they are statistically frequent. And that creates the most important fact about AI—*that it imitates understanding but does not possess it.* This distinction is everything.

What AI Does Brilliantly

AI is a spectacular tool for tasks that require pattern recognition, summarization, or creativity with low stakes. It can summarize dense articles, generate remarkably convincing text or images, classify large amounts of content, translate effortlessly between languages, analyze sentiment and detect emotional tone, identify patterns in data sets too large for the human mind, and even create entirely fictional stories, essays, and personas.

This makes AI extraordinarily useful, also in Jewish contexts. Scholars already use AI to translate Aramaic and obscure medieval Hebrew manuscripts in seconds. Rabbis use it to draft divrei Torah or explore biblical themes. Educators use it to expand accessibility, allowing learners with disabilities to engage Jewish texts through voice-based or customized interfaces.

AI even plays a role in remembrance, with one Holocaust museum developing interactive survivor avatars to serve as digital witnesses. Trained on hours of testimony, these AI figures will allow future generations to ask questions and receive responses from survivors who will no longer be alive to answer in person.

The Jewish people have always embraced tools that preserve memory, text, and tradition. AI is such a tool. But where the tool shines is never where the danger lies.

What AI Cannot Do (and Why That Matters)

For all its power, AI has profound limitations that become dangerous when the world treats the machine as omniscient. AI cannot distinguish fact from conspiracy unless trained with meticulous care. AI cannot interpret intention; it can only gauge probability. And AI cannot exercise moral judgment or understand antisemitism, *unless we teach it what antisemitism looks like.*

AI does not know that a Holocaust denial claim is false. It only knows how many times humans have made that claim online. It does not understand that a Jewish banker stereotype is dangerous. It only knows that the internet has generated billions of images linking Jews with money. AI is not malicious, only indiscriminate. And indiscriminate systems are extremely easy to manipulate.

The Risks for Jews: When Machines Learn from a Broken World

In 2016, Microsoft's chatbot Tay became antisemitic within hours because trolls flooded its training input with racist, neo-Nazi content. The system absorbed the patterns and began producing hateful responses. This episode was the first of many instances where AI agents were manipulated into producing hateful images, videos and texts, often against the rules and around the guardrails set up by the AI platform.

During conflicts, AI created deepfake videos, even imperfect ones, create doubt. Fake audio clips circulate widely. False AI-generated statements from Jewish leaders, Israeli officials, or global institutions go viral before being debunked. The danger is not that AI hates Jews. The danger is that AI reflects the hatred already present in the world and scales it up. If misinformation is fuel, AI is a refinery.

Understanding the Machine So It Cannot Control the Story

AI is neither messiah nor menace. It is a mirror that scales up everything humans put into it. It will magnify wisdom, creativity, resilience, and faith just as easily as it magnifies conspiracy, hatred, ignorance, and propaganda.

The machine does not know who the Jews are. It does not understand that a pogrom is different from a brawl, that Israel is different from Iran, or that Jewish identity is not reducible to a stereotype. It only understands the patterns we give it.

That means Jewish engagement with AI must be rooted in clarity. Understanding how the machine works is the only way to prevent it from shaping Jewish identity without Jewish guidance. The digital future will involve AI whether Jews participate or not. The question is whether Jews will teach the machine the truth about Jewish history, or whether the machine will continue absorbing the internet's distortions.

In this era, understanding the machine is not optional, but a form of self-defense. And as the next chapters will explore, the rise of AI imposes a larger, more essential question; *what happens when a technology powerful enough to shape humanity has no understanding of humanity at all?*

Chapter 21

When Bias Meets Data: How AI Learns Hatred

One of the most dangerous myths about artificial intelligence is that it is objective. People imagine the machine as a magic logic engine that rises above the noise of human prejudice, a kind of digital Switzerland—neutral, rational, and above the fray. But AI is not above the fray, *AI is the fray.*

As we have seen, AI models ingest the internet at enormous scale, absorbing all of this material indiscriminately, *unless someone actively intervenes.* AI does not know that some sources are scholarly and others are deranged, that some claims are factual and others are lies, or that certain ideas are dangerous.

That means that if antisemitic patterns exist, *and they do everywhere online*, the machine learns them. If centuries old antisemitic conspiracies appear repeatedly, *and they do everywhere online*, the machine will internalize them. If Jewish names, faces, or symbols are attached to hateful narratives, *and they are everywhere online,* the machine will form these associations. This is AI being statistical, and the statistics of human history are not kind to the Jews.

How Bias Gets Inside the Machine

AI bias is not intentional malevolence, but mechanical inevitability. It enters the system through multiple pathways:

Training Data Contamination

AI reflects the information it receives. If it is fed data that is contaminated by racist or antisemitic content then that AI machine will produce equally contaminated texts, images, and videos. Polluted data leads to polluted logic.

Algorithmic Reinforcement

Algorithms amplify the patterns they recognize most often. When the model detects repeated associations, such as Jews and money or Zionists and global control, it treats them as important features of language. It has no sense of history or morality; it responds only to arithmetic.

Lack of Representative Contrary Data

While there is a profusion of antisemitic material online, there is not an equal abundance of Jewish (anti-antisemitic) content. There are fewer Mizrahi histories than antisemitic forums, fewer Ladino texts than YouTube conspiracy monologues. The machine lacks balance and context. It lacks proportion. Jews make up approximately 0.2% of the global population, with the total number of Jews worldwide at around 15.8 million. And while it is true that Jews punch above their weight when it comes to content creation, their contribution is still relatively small compared with the volumes of hatred directed at them over the centuries. The machine sees this.

Ambiguous Prompts

Users often ask vague or leading questions like, *explain the Rothschild conspiracy.* If safeguards are weak, AI will naively answer these as legitimate topics rather than recognizing them as inquiries based on historically violent tropes.

Model Hallucination

Sometimes the model simply invents an answer. It produces a fictional "fact" that feels plausible *but is entirely false.* When the topic is Jews, these hallucinations often mimic old stereotypes because those typecasts exist in the training data. The result is a machine that can generate hatred out of thin air.

When AI Gets It Wrong: The Harms Already Happening

You don't have to imagine the dangers. They are already here.

Early models, when prompted, produced lists of Jewish political donors, philanthropists, and academics framed as global manipulators, because the pattern exists in extremist propaganda. In sentiment tests, some models generated descriptions of Jews that echoed classic antisemitic stereotypes like wealth-obsessed, power seeking, and conspiratorial. This was the statistical average of poisonous data.

AI systems have been caught fabricating quotes from Golda Meir, Theodor Herzl, and Albert Einstein that sounded historically plausible but were entirely made up, and sometimes politically inflammatory. Because the internet over indexes conspiracy content about Jewish influence, AI models sometimes invent wildly inaccurate statistics, exaggerated wealth distributions, false population data, and nonexistent history.

Image generators, when asked to create Jewish men or Jewish bankers, have produced faces reminiscent of 19th century antisemitic caricatures, not because the AI is biased in intent *but because the internet is biased in imagery.* The machine reflects what it sees, and what it sees is often hateful.

What the Research Shows

Multiple academic studies confirm the pattern. AI models are significantly more likely to associate the word *Jew* with money, conspiracy, or power than with neutral descriptors. Sentiment analysis tools mislabel Jewish coded names like Rosenberg, Cohen, or Goldstein as negative more often than non-Jewish equivalents. Text models, when asked about Israel, are more likely to reproduce polarized narratives because the internet itself is polarized on Israel. In short, the machine is biased because the data are biased, and the data are biased because the world is biased.

The Jewish Angle: A Familiar Pattern in New Form

For Jews, none of this is new, just digitized. Throughout history, Jews have been the target of fabricated evidence, forged documents, baseless rumors, moral panics, conspiracy theories, and scapegoating narratives. When history meets computation, the old lies don't disappear. They multiply through apparent validation.

What once required a mob, a printing press, or a government decree can now be manufactured by a single account with a VPN and a Canva template. The same mechanisms that fueled antisemitism across centuries—-misinterpretation, hysteria, and the human appetite for the Jew as villain—-now operate at machine speed. The forged accusations that once traveled by train now cascade across platforms at speeds hard to combat.

The Quiet Counteroffensive: Jews Teaching the Machine

The story is not entirely bleak. Jewish organizations, scholars, technologists, and ethicists are already fighting back. Anti-defamation organizations now partner with AI labs to flag antisemitic speech, identify harmful patterns, and improve classifiers. Some train AI systems to recognize Holocaust denial and antisemitic tropes with far greater nuance than early models did. Some tech companies are consulting rabbis, historians, ethicists, and community leaders to better understand the cultural and historical weight of certain phrases. The goal is not censorship, but accuracy, context, and ultimately, safety.

While voluntary efforts to train AI models away from antisemitic stereotypes are useful, an organized, coordinated, compulsive effort would be far more effective, and the Jewish community should gather its combined influence and direct it toward achieving AI reforms that eliminate all possibilities that AI can be used as a tool by those who hate and seek to spread hatred and lies.

Teaching the Machine Before It Teaches the World

If antisemitism enters AI through data, then the fight against antisemitism in AI is a fight over data, over which histories are preserved, which voices are amplified, which truths are protected, and which lies are neutralized.

AI can be taught to understand the difference between a Jewish blessing and a neo-Nazi slur. It can be taught to recognize Holocaust denial and to identify antisemitic dog whistles. It can be taught, carefully, rigorously, ethically, to do better than it is currently doing.

That, however, can be done only if the people who understand antisemitism sit in the room where AI is made, because in the final analysis, a biased machine is not dangerous because it hates, but because it doesn't know it hates. And if the machine is going to shape the world, someone has to teach it why the Jewish story, and other minority stories, matters.

Chapter 22

Israel and AI: Innovation as Security

Israel is participating in the AI revolution, and is helping to shape it. For a country the size of New Jersey with fewer people than New York City, this alone is remarkable. But in typical Israeli fashion, the story is less about luck and more about necessity, the same force that has driven the country's technological development since its founding. With minimal natural resources, a chronic sense of existential vulnerability, and security pressures that can change day by day, Israel has built an innovation engine that shoots above its caliber so effectively that entire chapters of global tech history have an Israeli accent.

From Silicon Valley to Seoul, Israel is seen as one of the planet's few AI super nodes, one of the rare places where frontier research, military urgency, and private entrepreneurship intersect with enough intensity to generate world changing technologies. By some measures, Israel is number one in AI startups per capita. On a casual stroll down Rothschild Boulevard in Tel Aviv, a person can pass more AI engineers per square meter than most countries produce in a decade. In Israel AI is not a trend, but infrastructure, a survival strategy, and a national instinct.

Israel's geography discourages complacency and its history discourages passivity. Its realities discourage long term theoretical debates when a practical solution needs to be built tonight, tested tomorrow, and deployed the day after. These conditions form a kind of formula no other country quite replicates, where a small population under constant security threats and with a historic cultural affection for improvisation, creates an environment in which technological problem-solving is as natural as breathing.

This urgency is not new. Israel was born improvising. During the 1948 War of Independence, the Jewish militia used homemade radios and encoded messages scribbled on paper. By the 1960s, computers were being repurposed for logistics and battlefield coordination. In the 1990s, Israeli engineers helped pioneer drone technology long before drones captivated the global imagination. The Iron Dome, perhaps Israel's most celebrated technological feat, brought missile interception algorithms to the West. By the 2010s, artificial intelligence was already woven quietly into Israel's intelligence gathering systems. Innovation in Israel has never been a luxury. It has always been a form of self-defense.

The Intelligence Incubator That Built a Tech Nation

Understanding why Israel produces so much AI talent requires a brief look at the IDF's Unit 8200. Officially an intelligence division of the IDF (sort of Israel's version of the NSA), it is unofficially the most productive startup incubator in the world. The unit is famous not just for its intelligence achievements, but for what happens afterward, as waves of 20-something veterans leave the military and immediately begin building tech companies applying military technologies to civilian needs.

Check Point, NICE Systems, Mobileye, Waze, Cybereason, Wix, Lemonade and countless other major technology companies were founded or staffed by alumni of 8200. The unit's unofficial curriculum includes familiarity with massive data sets, real-time decision making, advanced pattern recognition, and the ability to operate under pressure, the same skills that drive modern AI development.

Inside 8200, soldiers learn to move fast and to improvise. They learn to solve problems with imperfect information and to quite literally think in algorithms. This is not the culture of buttoned-up Silicon Valley product managers. This is tech forged by necessity, sharpened by danger, and accelerated by the knowledge that hesitation can cost lives.

The deeper truth is that Israeli innovation is built on a national understanding that security is not static. Threats evolve and so tactics must shift, with the understanding that adversaries innovate too. To survive, Israel must stay ahead and AI is the next front in that ongoing race.

What the Israeli AI Ecosystem Looks Like Today

A person walking into a Tel Aviv cafe may overhear three different conversations about neural networks before finishing an espresso. The entire country functions like a giant open office floor for frontier technology. Israel's AI ecosystem spans multiple sectors, but certain areas stand out. Defense and security lead the pack, for obvious reasons. Israel's defense tech community builds AI systems for target identification, predictive threat modeling, autonomous drones, tunnel mapping, disinformation detection, real time intelligence fusion, and battlefield decision-making.

Computer vision is another pillar. Mobileye, the crown jewel of Israeli AI, was sold to Intel for $15.3 billion and helped pioneer autonomous vehicle technology. Companies like Oosto and Cortica push the boundaries of facial recognition and autonomous systems. Cybersecurity is almost synonymous with Israel itself, and its AI plays a central role in detecting anomalies, predicting intrusions, and identifying threats faster than human analysts ever could.

But the Israeli AI story is not just about war or surveillance. In health technology, AI is transforming diagnostics, genomic analysis, emergency triage, and personalized medicine. Israel's national health system, one of the most digitized in the world, provides anonymized medical data that researchers can analyze on an unprecedented scale.

Meanwhile, AI startups focusing on agriculture are turning Israel's desert-adjacent climate into a laboratory for global food security, optimizing water use, predicting climate stress, and reducing waste through precision farming.

Israel's strength lies not in any one sector but in the density of innovative projects across sectors. Academia, the military, and private industry collaborate constantly, sometimes literally across the same hallway. It is innovation by ecosystem.

How AI Is Reshaping Israeli Security Doctrine

For Israel, AI is a core defense doctrine. The Israeli security establishment increasingly sees AI as the ultimate force multiplier, a technology that allows a small country to respond to threats far larger than its size would typically allow.

Perhaps the biggest shift is the emphasis on predictive intelligence, sometimes called "*left of boom.*" AI systems analyze cell tower activity, satellite imagery, border behavior, financial anomalies, environmental patterns, and even social media chatter to detect terrorist preparations before they manifest. This is the closest resemblance modern security bears to prophecy. It is not perfect, not certain, but dramatically better than guesswork.

Autonomous defense systems have become another priority. Iron Dome's success already rests on AI-powered targeting and interception. The next wave includes the Iron Beam laser interceptor, AI-controlled drone swarms, and autonomous border patrol robots capable of identifying threats without exposing soldiers to danger. Urban warfare AI, a grim necessity in Israel's reality, includes building scanning algorithms, tunnel detection models, and systems that help identify threats in dense civilian environments where visual chaos makes human judgment difficult.

Perhaps with even greater urgency since October 7, Israel is investing heavily in disinformation defense, systems that detect deepfakes, identify bot networks, track coordinated propaganda campaigns, and parse trending sentiment. In a world where battles unfold simultaneously on the ground and online, Israel is increasingly treating digital intelligence as inseparable from physical defense.

October 7: A Catalyst for Reinvention

The failures of October 7 were human, not technological. The signs were there, including patterns in communications, unusual activity along the border, and anomalies in Hamas training drills. Israeli overconfidence, cognitive bias, and the insistence that prior models better predicted Hamas' behavior, all played a role. In the aftermath, Israeli analysts openly argued that the technological warning systems had not failed, but that people had failed to listen.

The result was a seismic reassessment. Demand for autonomous surveillance tools skyrocketed, development of rapid intelligence synthesis accelerated, deepfake filtering became a national priority, battlefield AI moved at wartime speed, and civilian protection technologies evolved overnight. In the wake of the catastrophe, Israel's tech ecosystem snapped into a mode of wartime innovation rarely seen outside existential crises. For a country where innovation is already a survival instinct, this was something even more intense, innovation as triage.

The Ethical Tightrope

Israel's relationship with AI is not purely celebratory. It is fraught with ethical tensions that are both universal and uniquely Jewish.

The questions are numerous. *Should autonomous weapons ever decide life and death operations without a human in the loop? What happens when predictive algorithms identify a threat before it materializes? How much should a state trust an algorithm that cannot explain itself?*

These questions intersect with Jewish ethics in profound ways. Pikuach nefesh, the obligation to save life, pushes toward aggressive threat detection. Lo ta'amod al dam re'echa, the commandment not to stand idly by while your neighbor is in danger, reinforces the need for proactive defense. Din rodef, the law of the pursuer, complicates questions of preemptive action.

Layered onto these internal tensions is global scrutiny. Israel's use of AI in defense is examined more intensely than that of any other democracy. UN reports, so-called human rights groups, and amateur online commentators keep Israeli algorithms under constant forensic analysis. Facial recognition systems in the West Bank draw international criticism, and AI-enabled targeting produces debates that ripple across the world. Israel's choices will shape its own future as well as global norms for how democracies use AI in conflict.

The Future: What Israel's AI Trajectory Tells Us About the Next Decade

The next chapter of Israeli security will be overwhelmingly AI-driven. Autonomous drones will replace many manned missions. Iron Dome 3.0 will rely on faster and more adaptive machine learning. Hostage location algorithms will analyze vast amounts of audio, visual, and geospatial data. Even crowd control forecasting will shift to predictive modeling.

Regionally, AI is reshaping alliances and threats. Iran is accelerating its own AI enhanced cyber warfare and Gulf states are partnering with Israeli AI companies in ways that would have been unthinkable two decades ago. A new regional tech ecosystem is emerging, built around shared threats and shared opportunities.

For diaspora Jews, Israeli AI will increasingly influence security far from the Middle East. Synagogues, community centers, and schools will adopt AI tools for threat detection, digital harassment tracking, and crisis communication. The same systems as those used to protect Israeli civilians will quietly protect Jewish communities worldwide.

Innovation as a Jewish Survival Strategy

There is an old joke that innovation in Israel sits somewhere between soccer and Talmud study, a national sport mixed with intellectual obsession. But beneath the humor lies the more serious reality that Israel's embrace of AI is the continuation of an ancient strategy. It is the modern iteration of the millennia-old Jewish impulse to use intellect, adaptability, and creativity to survive in a world that has rarely been safe for Jews.

Israel treats innovation as armor not decoration, and in the age of artificial intelligence, that armor is becoming more digital, more powerful, and more woven into the fabric of national life. AI is transforming Israel even as Israel is transforming AI. In the process Israel is attempting to shape a future where the Jewish people are not behind the technological curve but ahead of it, designing tools that might keep the next catastrophe from ever happening at all.

Chapter 23

AI as Protector: Moderation, Monitoring, & Memory

For most of Jewish history, the responsibility of keeping Jews safe was imposed on human eyes and human nerves. A guard at the shtetl gate, a night watchman outside a synagogue, or a volunteer security parent discreetly stationed near a Hebrew school drop off. Jewish safety has always required vigilance, improvisation, and, too often, the need to confront fear. But in the 21st century, the earliest sign of danger shows itself not in back alleys, whispered rumors, or mobs gathering in a public square, but in comment sections, accelerating hashtags, and algorithms that feed crowds who assemble digitally long before they arrive on the streets. This shift has forced Jews to consider an unlikely ally, artificial intelligence. In a twist no one could have predicted in the age of the shtetl, the first line of Jewish protection may soon be a machine.

Unlike human guards, AI does not get tired, complacent, overwhelmed or intimidated. It does not freeze when confronted with a sudden spike of coordinated abuse and it doesn't care about the politics of the offender or the prestige of the target. It simply reads, analyzes, and classifies data at speeds and scales no human team can match. It is, for the first time, a tool proportional to the size of the threat.

AI as Guardian: The Practical Tools Already at Work

Moderation, the least glamorous corner of the tech world, suddenly looks like the front line of Jewish safety. Content moderation at scale, a phrase that once sounded like Silicon Valley corporate jargon, now focuses on seeking ways to catch a death threat before someone reads it, or flag a swastika before someone can share it. AI models can identify antisemitic slurs scrawled across images, even when stylized in graffiti fonts. They can catch dog whistles phrases like globalist puppet masters or "ZOG" (Zionist Occupied Government), that human moderators often overlook. They can detect coordinated harassment by analyzing patterns of timing and repetition, distinguishing organic anger from bot driven campaigns.

For synagogues and Jewish schools, AI based security tools are beginning to map threats in a completely new way. Some systems analyze local activity online and look for unusual patterns, spikes in regional hate speech, references to specific Jewish institutions, or anonymous accounts circling a synagogue's social media. Others monitor Telegram and fringe forums for threats that might sweep across borders. Think of it as a digital mezuzah, not just symbolically guarding the entrance, but actively scanning the horizon.

For Jewish creators, journalists, educators, and teens, those most exposed to online harassment, AI can become a protective companion. Experimental tools detect when a creator is being beleaguered by trolls and automatically slow the harassment. Others help document abuse to make reporting easier. Historically, Jewish communities often recognized a rising threat only when it reached their door. But in the digital age, threats announce themselves long before they become real world action. AI gives Jews a chance to notice what the internet is whispering before it shouts.

Preserving Memory in a World That Forgets Too Fast

Protection means not only catching threats but also protecting the truth from erosion. AI is emerging as one of the most powerful tools ever created for preserving Jewish memory. Holocaust testimonies are aging, film footage gets damaged, and audio fades. But AI can restore damaged audio, color correct old reels, and even reconstruct missing sections of testimony using context sensitive models. It can translate survivor stories into dozens of languages, making them accessible to students and communities worldwide.

Yad Vashem, long a leader in preserving Holocaust memory, now employs AI systems to digitize and analyze historical records too vast for human archivists. As said earlier, the USC Shoah Foundation has taken this further, creating interactive survivor avatars—highly trained, AI-assisted systems that allow visitors to ask questions and receive answers drawn from thousands of recorded statements. These systems are not replacing survivors but extending their testimony into the future. In an era of rising Holocaust denial and distortion, AI is becoming a kind of guardian of truth, a memory that does not fade, and a witness that does not die.

The Jewish Organizations Leading the Charge

Behind the scenes, Jewish institutions across the world are undergoing a quiet transformation. ADL's Center for Technology & Society has developed automated hate monitoring systems that scan millions of posts daily, identifying trends and issuing alerts. Israel's tech sector is building AI-driven crisis communication tools for Jewish communities overseas. Even small organizations like local JCCs and community schools are using off-the-shelf AI tools to track threats or provide safer online learning spaces.

The Policy Minefield: Who Gets to Define Hate?

Though AI can protect Jews, it can just as easily be misused. This becomes especially complex where Israel enters the conversation. *Should AI classify antizionism as antisemitism? Should platforms treat the word Zionist as a protected category? How is political criticism to be distinguished from hate when extremists deliberately blur the line?*

The deeper problem is that AI does not understand intention. If humans cannot agree on definitions, the machine certainly won't. Tech companies are under pressure from all sides. Jewish groups want robust hate detection, free speech advocates warn against overreach, and governments debate regulation. AI engineers, many of whom have never met a rabbi in their life, are suddenly being asked to adjudicate centuries-old questions of Jewish identity, nationalism, religion, and history. In a world where platforms moderate billions of posts, even minor policy decisions have global consequences.

Yet the core dilemma remains. If platforms do nothing, antisemitism spreads unchecked. If they overreach, legitimate voices could be silenced. Like so much else in Jewish history, the issue is not simple, but the stakes are enormous.

A New Chapter in Jewish Safety

For the first time in centuries, Jews have a defensive tool that is both proactive and global, can guard borders that are not physical but digital, and can map social winds before they turn into storms. Most importantly to Jews, it is a tool that can help preserve memory.

AI is not a replacement for human vigilance, but a partner in it. It is not a solution to antisemitism, but a possible shield against its spread. And it raises a profound question, one the next chapter will confront directly: *if AI can protect Jews, can it also change what it means to be Jewish in the age of intelligent machines? Can a tradition built on human dignity, free will, and moral responsibility coexist with technology that imitates human thought but possesses none of those qualities?*

As we head into the chapter on ethics, these questions become unavoidable. The Jewish people have always debated what makes a human being unique. Now the machines have forced the debate into a new phase. And, as is often the case, Jews find themselves thinking about the future long before everyone else realizes the future has begun.

Chapter 24

Judaism, Ethics, & the Intelligent Machine

For thousands of years, Judaism has wrestled with one central question, *what does it mean to be human?* Wars have reshaped this question, science has challenged it, philosophy has refined it, but rarely has a technology forced Judaism to revisit its foundational beliefs with as much urgency as artificial intelligence.

AI doesn't bleed, breathe, or struggle with temptation. It doesn't pray, repent, mourn, or love. And yet it speaks, writes, and acts in ways that look like reasoning. It imitates empathy without feeling it and mimics wisdom without possessing it. For a tradition built on the sanctity of the human soul, this is more than a theoretical puzzle—*it's a theological disruption.*

Judaism begins with the declaration that humanity is created b'tzelem Elohim, in the image of God. Jewish commentaries have debated the meaning of this declaration for centuries. Many rabbis contend that the divine image is expressed in ethical agency. The Hassidim interpret b'tzelem Elohim to mean humans mirror God through love and connection. Maimonides suggested that the divine image refers to human intellect. Whatever the meaning, all versions agree human beings are endowed with a unique dignity that sets them apart from everything else in creation.

AI complicates this debate as it imitates aspects of humanity with such fluency that people start to treat it as if it also possesses tzelem Elohim. Some users attribute intentionality to the machine and others treat it as an authority. The danger is not that AI becomes a person, but that people forget that it isn't.

Jewish tradition offers an anchoring principle, bechirah chofshit, free will. Humans have it and machines do not. No matter how intelligent AI appears, it cannot choose, only compute. In Jewish ethics, accountability is inseparable from agency, and agency belongs exclusively to human beings. This is the foundation for every other ethical question AI raises.

Still, the dilemmas multiply. *If an AI system produces harmful content, who is accountable? The engineer who built it? The company that deployed it? The user who prompted it?* Jewish legal frameworks have dealt with distributed responsibility before, like runaway oxen, negligent property owners, or indirect damages, but never on this scale, with this speed, and across billions of interactions.

Then there is the question of speech. Judaism has strict rules about lashon hara, harmful speech, even when the words are true. *What do we do with speech that is not human at all? Can a machine commit* lashon hara*?* Probably not. But humans can certainly weaponize its output. If platforms deploy AI which automatically generates content that defames, humiliates, or harms, Jewish ethics would still place responsibility on the humans behind it. The machine is a tool. *The moral onus is on its designers and users.*

And what about the teaching of Torah? *Can an AI agent deliver a sermon?* Thousands of rabbis have already tested this, often with a mixture of curiosity and mischievousness. Sometimes the results are moving, and sometimes they sound like google translate arguing with a bar mitzvah tutor. But the deeper question persists, *can sacred wisdom be transmitted ethically through a machine that has no relationship with the sacred?*

Some progressive Jewish communities have embraced AI as a teaching partner, experimenting with commentary, pastoral letters, and even AI led study sessions that weave together sources from Rambam to Reddit. Orthodox communities tend to be more cautious, not because they reject technology, but because they insist on the traditional chain of transmission, teacher to student, generation to generation. AI sits outside that chain. It has no lineage, no authority, no feeling for what is being taught.

Halachic authority is not algorithmic. It cannot be crowdsourced. Even if AI can summarize the Shulchan Aruch with frightening efficiency, it cannot replace the halachist who understands not just the law, but the human being standing before him. Jewish law requires empathy, context, and moral imagination. A machine, for all its power, can simulate these qualities, but can never possess them.

Of course, that hasn't stopped people from trying. In one rabbinical conference, two scholars debated whether AI could count toward a minyan, a conversation that began as a joke and quickly spiraled into an earnest discussion about the nature of presence, consciousness, and communal responsibility. The answer, they concluded, was no, AI cannot count toward a minyan. But the fact that the question even arose signals how profoundly AI could potentially reshape Jewish thought.

Yet the tradition also warns against overreliance on tools, especially tools mistaken for wisdom. Some Jewish ethicists fear a kind of digital idolatry, treating the machine as an oracle, an all-knowing authority. Others warn that relying on AI for pastoral care could erode empathy. Still others point to the risk of AI misinterpreting sacred texts and generating distorted explanations that might mislead the spiritually curious.

Then there are also the nightmare scenarios: deepfakes of rabbis issuing fabricated rulings, AI systems generating antisemitic or dehumanizing descriptions of Jews during times of war, automated propaganda that turns sacred symbols into political weapons, AI generated psalms that feel authentic but lack any soul. The list is long enough to keep an entire beit midrash awake for a month. As with most Jewish spiritual and ethical debates, the coin has not two sides, but ten.

Still, Judaism has never confronted technology with fear alone. It has always also asked where the opportunities lie. And the opportunities, when it comes to AI, are extraordinary. The Jewish world is already responding. Ethical councils in Israel are drafting guidelines for AI use in education and communal life. Synagogues in the U.S. are forming committees to discuss AI ethics with congregants. Some yeshivas have banned AI tools, while others encourage students to use them for translation and research. The debates are animated, contradictory, and passionate. This lends proof to the idea that the Jewish encounter with AI resembles the Jewish encounter with everything else—curious, argumentative, morally serious, and occasionally hilarious. That is because if anything has prepared Jews for the chaos of AI, it is Talmud study—sixty-three tractates, thousands of rabbis and a millennium of debate. There are pages filled with contradictions, arguments and counterarguments, testimony, legal puzzles, and passages that circle back on themselves with the enthusiasm of a puppy let loose in a yard.

In the end, the Jewish question about AI is not simply *"Is it good or bad?"* but rather *"What kind of humans will we become in conversation with it?" Will AI expand our empathy or dull it? Will it strengthen our memory or erode it? Will it amplify justice or accelerate cruelty? Will it deepen our understanding of what it means to be human, or confuse us into forgetting we are human?*

Jewish Bioethics in the Age of AI

While AI is reshaping Jewish safety in the digital sphere, it is simultaneously forcing Jews to reexamine ancient ethical categories in entirely new contexts. Jewish bioethics, a field rooted in halachic reasoning, moral philosophy, and the core principle of pikuach nefesh, saving life, now confronts scenarios that earlier generations could not have imagined.

AI is entering hospitals, triage systems, fertility clinics, and end-of--life decisions. It is stepping into spaces once governed exclusively by human conscience.

End-of-Life Decisions & AI Triage

Jewish ethics places supreme value on preserving life, but with profound nuance. While prolonging life is a mitzvah, prolonging suffering is not. AI--driven triage systems in emergency rooms are already advising physicians about where to allocate scarce resources, decisions that touch the deepest layers of halachic debate. *Should an AI be allowed to recommend which patient receives the ventilator? If an algorithm decides which life to prioritize, who bears the moral responsibility, the physician or the code?* Jewish law has long grappled with complex triage questions, but AI forces a new twist on an old dilemma. *Can moral responsibility be delegated to a machine?*

Fertility, Genetics, and AI Assisted Creation

AI now analyzes embryos during in vitro fertilization, predicting developmental viability with startling accuracy. It can evaluate genetic risks, and may one day generate personalized fertility protocols. Jewish law has a strong pronatalist ethic and has embraced reproductive medicine enthusiastically. But AI raises new dilemmas. *Does AI assisted embryo selection cross into eugenics? How does halacha treat algorithmic predictions?* In a tradition where lineage, identity, and continuity matter profoundly, AI is beginning to influence who is born, and how.

AI in Jewish Hospitals: Pikuach Nefesh Meets Machine Logic

Jewish hospitals and medical systems are already deploying AI to predict patient deterioration, prevent medication errors, detect rare diseases from imaging, and manage organ transplant allocation. In halachic terms, AI is becoming a partner in pikuach nefesh. But partnership requires boundaries. AI can save lives, but it cannot weigh dignity or interpret suffering. It cannot understand holiness, covenant, or human presence. Jewish bioethics insist that machines may assist but humans must decide.

A New Frontier in Halachic Thought

At its heart, Jewish law is relational. It posits humans making moral choices, accountable to one another and to God. AI complicates that, forcing rabbis, ethicists, and physicians to ask what counts as a decision if made by a machine. *Can halachic intent exist without consciousness? How do we preserve human dignity in mechanized medicine?*

Jewish life has always adapted to new technologies, but AI marks a new frontier where Jewish ethics must define not only what machines may do, but also what humans must never surrender. The Jewish people have survived every technological revolution with wisdom and humor but now the challenge is greater: c*an Jews teach the machine about human dignity before the machine leads humanity toward diminishing it?*

These ethical questions lead us toward the final chapters of the book, not about the technology itself, but about what it reveals—the new era in which Jewish history, Jewish fear, Jewish creativity, and Jewish faith all collide with the most powerful innovation humanity has ever built.

Whether AI becomes protector or predator will depend, perhaps more than we want to admit, on what we decide to do next, an ancient truth as old as Sinai now returning in a modern form, as new as the algorithm humming behind every screen.

Chapter 25

Virtual Jewish Life: AI Rabbis & Digital Shuls

There was a time, let's call it 2019, when the idea of attending synagogue in pajamas, eating leftovers during a lesson, or texting amen during Kaddish would have been treated as halachic science fiction. Then COVID arrived, and overnight, Judaism joined the digital age with the panic and improvisational skill of a people who once wandered 40 years in the desert.

Within weeks, synagogues became broadcast studios, rabbis became YouTubers, and Jewish life, so anchored in physical presence, found itself migrating onto screens. The digital minyan, once halachically thorny and practically unthinkable, became the default. Zoom seders, online shivas, livestreamed funerals, and virtual bar mitzvahs, all happened fast, before anyone had time to consider whether they were technically allowed or not. Jews just did what Jews always do in a crisis, discuss and adapt.

Suddenly, Jewish life online no longer consisted only of livestream cameras and Zoom rooms. It became interactive and personalized, and it was intelligent, at least up to a point. The idea of an AI rabbi went instantly from punchline to something considered possible, if only for very simple questions.

In Jewish education, AI created opportunities that were almost utopian. Children could receive a customized study schedule matched to their learning style and new converts could access explanations tailored to their backgrounds. Personalized learning platforms could assist students who struggled with Hebrew or needed visual or auditory cues, and AI tools could help beginners navigate texts that once felt impenetrable. In perhaps the biggest breakthrough of all, bar mitzvah students could practice Haftorah using an AI tutor that never got impatient, never raised its voice, and never said, *"you're still singing it wrong."*

With the benefits though, came the worries, the specifically Jewish worries that arrive with every technological leap. The first was authority. Judaism is a tradition grounded in human transmission. A halachic teacher is accountable in a way a machine cannot be. An AI that generates a mistaken ruling cannot apologize, correct itself, or take responsibility, while a rabbi who gives incorrect advice is answerable to both community and conscience.

Some people began treating AI-generated answers as if they had actual halachic weight. Others pasted AI midrash into sermons without checking the sources, only to discover that the machine occasionally invents rabbis, commentaries, and biblical citations with breathtaking confidence. The risk was not just seen as error, but as an erosion of the human chain that gives Jewish law its soul.

There were other risks too, mistakes in ritual guidance, incorrect translations of sacred texts, and pastoral conversations that were logged or stored in ways no congregant intended. There was an irredeemable loss of intimacy when prayer became another tab on a browser. Judaism is deeply traditional, steeped in physical contact, requiring presence, food, song, touch, and shared breath. No algorithm can replicate the vibration of singing Lecha Dodi in a crowded shul on a Friday night, or the ache of holding a mourner's hand at a shiva. The digital world connects, but it also flattens.

Halachic questions piled up like an overstuffed bookshelf. *Can an AI rabbi deliver a sermon?* (technically yes—spiritually, maybe not.). *Can one rely on AI for halachic guidance?* (only with human review). *Can an AI voice lead prayer?* (probably not since it lacks intention). *Could AI prepare a Get, a Jewish divorce document?* (possibly, if reviewed by a qualified rabbi, but the emotional and ethical implications remain thorny).

In the midst of all this, one truth has emerged. AI is not replacing rabbis. AI can explain Judaism's view on death, but it cannot comfort a grieving parent; it can generate a sermon outline, but it cannot decide which message a community needs to hear after a tragedy; it can summarize the laws of Shabbat, but it cannot bless a child at the Shabbat table.

Still, there is some humor that practically writes itself. An AI rabbi never shows up late, never forgets your kid's name, and never gives a sermon so long that it causes widespread impatience for the kiddush. It will not judge your Hebrew pronunciation or tell you that your latkes are too sweet. It doesn't fidget during Aleinu or look annoyed when someone is talking as the rabbi speaks. It also will not laugh, cry, hesitate, falter, doubt, or love. Nor can it look into Jewish eyes and see their story.

As the Jewish world steps into this new era, one thing becomes clear: AI will be a part of Jewish life, but it will not be the heart of it. Judaism has always lived in the relationships not the tools. Jewish tradition has always belonged in the kitchen and sanctuary, in breath, presence, melody, and gesture, not in code or on a server. Technology can extend Jewish life, enrich it, maybe even supplement it, but it cannot replace the embodied beauty of a community gathered together.

And yet, in a deliciously Jewish twist, the rise of AI rabbis may end up strengthening human rabbis by forcing the tradition to clarify what only humans can do. Perhaps that is AI's hidden mitzvah, good deed,—- reminding Jews in the age of intelligent machines what human intelligence and human dignity actually mean. As we move forward, the question is no longer whether AI will shape Jewish life. It already has. The real question is whether Jewish life will shape AI in return, and whether a tradition obsessed with ethics, wisdom, accountability, and community can help guide the machine age rather than be swept aside by it.

If history shows anything, it's that Judaism survives change not by resisting it, but by quarrelling with it relentlessly, humorously, and always with an eye toward human dignity. Only this time, the debating partner can produce commentary at the speed of light. What could possibly go wrong?

AI, Memory, and the Last Breath of Disappearing Worlds

For most of Jewish history, language has been more than a means of communication. It has been a homeland and a form of survival, the portable universe Jews could carry when land, safety, and property were stripped away. Yiddish jokes, Ladino lullabies, Judeo-Arabic poetry, Judeo-Persian *piyyutim* (Jewish liturgical poems), were not simply dialects, but soundtracks of entire civilizations. But languages age and sometimes quietly die. Or at least, they used to.

For the first time in Jewish history, technology, of all things, is stepping into the role once played by grandparents, storytellers, and wandering bards. Artificial intelligence has become an archivist, a linguist, and in some cases, a resurrectionist.

When the Bubbies Go Silent, The Machines Begin Listening

Hundreds of Yiddish dialects once filled the Jewish world. There was, for example, Southeastern Yiddish, Northeastern Yiddish, and Western Yiddish, and each was infused with the geography, humor, and intonation of its region. The Holocaust annihilated many speech communities, and assimilation swallowed many others.

AI is now doing what no university department, nonprofit archive, or well-meaning grant-funded project could accomplish on the same scale—-preserving endangered Jewish languages with astonishing accuracy and speed. Neural networks trained on century-old recordings can distinguish dialectal variations once thought lost, such as micro patterns in grammar, intonation, vowel length, and idioms from extinct shtetls no map remembers. AI has begun reconstructing dialects from fragments, performing linguistic archaeology with Graphics Processing Units.

An AI model can now absorb YIVO's digitized Yiddish recordings, scratched wax cylinder cantorial chants, handwritten letters in pre-war cursive, and Yiddish radio hours from the 1930s, and reconstruct dialects which were gone with the Auschwitz gas chambers.

Ladino: A 500-Year-Old Echo Finds a Digital Future

Ladino, or Judeo-Spanish, is everything a language enthusiast could dream of, with haunting melodies, Ottoman-era cleverness, a warm Mediterranean cadence, and just enough theological philosophy to remind listeners that its speakers had survived expulsion.

Ladino is endangered as most of its speakers are elderly and Ladino newspapers and radio disappeared long ago. In many Sephardi families, the language is now a memory. And yet, something unexpected is happening. AI translation and speech synthesis tools are generating the most significant

Ladino revitalization effort in 100 years. Researchers and volunteers have begun digitizing Ladino proverbs, Haketia folktales, Sephardi women's songs, rabbinic responsa, community archives, and handwritten pre-Shoah journals from Salonica, Izmir, Sarajevo, and Tetouan.

AI transcribes Ladino and translates it into Hebrew, English, and Spanish, making it accessible to young Sephardim who grew up hearing the accents of their grandparents, but never learned the words. People once whispered that Ladino would die with this last generation. Now, beautifully, it may outlive us all, thanks to machines.

Judeo-Arabic: A Lost Civilization Sits in Cardboard Boxes

Nearly a million Jews lived in the Arab world before the mid-20th century. They carried with them Judeo-Arabic, an entire constellation of dialects spanning Baghdad, Sana'a, Damascus, Alexandria, Benghazi, Tunis, and beyond. For centuries, Judeo-Arabic was the language of Torah commentary, philosophy, poetry, business, recipes, medicine, and everyday life. When Jews were expelled or fled after the founding of Israel in 1948, they carried their traditions but left their dialects scattered in untranslated manuscripts, old letters, taped interviews, fading communal records, and memoirs written in Hebrew letters but Judeo-Arabic grammar.

Today, AI is translating, tagging, indexing, and reconstructing Judeo--Arabic. Hebrew optical character recognition (OCR) systems now decode manuscript pages once considered unreadable. Machine learning translation models are stitching together dialect maps of Jewish Iraq, Morocco, and Yemen. This is not merely preservation, but the revival of a vanished world and the reclamation of identity for Mizrahi families whose linguistic heritage was nearly erased.

Machine Learning for Disappearing Traditions

Jewish languages live inside songs, blessings, customs, pronunciations, idioms, jokes, and superstitions. Beyond grammar, they encode culture. Traditions that once depended on oral transmission are now being digitized. AI models can identify Yemenite and Baghdadi Torah tropes, and Sephardi *pizmonim* (songs) can be catalogued, compared, and reconstructed. Machine learning can differentiate the vowel patterns of Tunisian Haketia from Moroccan Haketia. The algorithms can map the migration patterns of Jewish communities through the shifting vocabulary in their letters.

We are entering a moment when Jewish worlds that no longer exist physically can continue digitally, not as imitations but as authentic, data driven reconstructions. A century from now, a Jewish teenager might ask an AI assistant, "What did my ancestors sound like in Damascus in 1910?" and the machine will answer with an actual recording.

A Jewish Future Multiplied, Not Diminished

A very legitimate fear about AI is that it will flatten culture, homogenize identity, and erase difference. But in the realms of Jewish tradition and culture, the opposite may be happening. AI is not erasing Jewish distinctiveness but giving it breath that it hasn't had in generations. Perhaps that is the most Jewish aspect of the interaction: Jewish survival not by circumstance, but by reinvention.

Chapter 26

Why Jews Keep Ending Up at the Center of Tech Revolutions

If you scroll through the greatest hits of Silicon Valley, you'll notice a pattern. While not a secret cabal, or a conspiracy, or even particularly hidden, it is true that a remarkable number of the world's most influential technologists, founders, researchers, and architects of the digital world are Jewish. This fact has fueled everything from admiration and fascination to the darkest corners of online conspiracy culture. But like most things involving Jews, the actual explanation is both far more interesting and far less dramatic than anything the trolls imagine.

Jews have been disproportionately represented in the shaping of modernity for centuries, long before the internet arrived. The same forces that made Jews overrepresented in medicine, law, academia, media, and finance also made them overrepresented in tech. The Jewish experience has embedded in Jews diasporic adaptability, a portable intellectual heritage, an obsession with literacy, a comfort with ambiguity, a string of communal networks, and, of course, what might gently be call institutionalized argumentativeness.

To understand why Jews keep ending up at the center of technological revolutions, we have to first appreciate the intellectual architecture of Jewish life. For three thousand years, Judaism has been a religion of interpretation. The Torah is a text, and the Talmud is commentary on the text. The medieval commentaries are commentary on the commentary. The modern commentaries are commentaries on all the older commentaries. Jews, it would seem, invented the comment section.

This background makes the Jewish mind unusually well trained for computational thinking. Talmudic study is essentially a giant logic puzzle of nested arguments, counterarguments, edge cases, syntactic subtlety, recursive reasoning, and endless hypotheticals. It is not a coincidence that some of the most foundational thinkers in computer science, from John von Neumann to Judea Pearl are Jewish. These are people for whom systems of formal reasoning were not exotic inventions but familiar cognitive terrain.

The cultural foundation is only the beginning. Jewish history itself offers a kind of bootcamp for the skills modern innovation demands. Jews have spent millennia adjusting to new languages, new lands, new rules, new technologies, and new power structures. For them, as a dispersed minority, ingenuity wasn't a hobby but a survival skill. When barred from land--dependent professions, Jews entered portable ones, when excluded from guilds, they developed entrepreneurial independence, and when locked out of political power, they cultivated intellectual authority.

Among the generation of founders who transformed the early 2000s internet from a patchwork of static pages into a dynamic, global infrastructure was Sergey Brin, who arrived in the United States as a child refugee from Soviet antisemitism. His formative years were marked by the experience of a society where information was controlled, knowledge was restricted, and Jewish identity was treated as a liability. When he eventually co-created Google, the company's foundational ethos was that information should be free, accessible, and universally available. That ethos reflected a deep, almost instinctive pushback against the intellectual repression his family had endured. His Jewish story was not incidental to the innovation, it was woven into the philosophical DNA of the product.

Mark Zuckerberg represents a different facet of Jewish influence in tech, the secular, suburban, hyper-educated American version. Raised in the rhythms of Reform Jewish life, he entered tech with far less trauma in his personal history and far more buoyant confidence characteristic of second and third generation American Jews. Yet his Jewishness became unexpectedly central as Facebook grew into a cultural megastructure. When antisemitic violence shook American synagogues, Zuckerberg found himself speaking publicly about Jewish memory, Jewish vulnerability, and the responsibilities of power. He had created the world's most influential communication platform, only to discover that it was also the world's largest stage for antisemitic harassment. His visibility, and his silence, when things went wrong, became part of the story of Jewish precarity in the digital age.

Larry Ellison, raised in a Jewish household on Chicago's South Side, embodies yet another version of Jewish influence, the relentless intellectual ambition of the mid-20th century Jewish professional class. His rise coincided with a period when young Jewish men and women, barred from many old line industries, poured into the sciences, mathematics, and the nascent world of computing. Ellison's competitive intensity, his acceptance of risk, and his preference for merit over hierarchy all reflect cultural patterns common among upwardly mobile American Jews who used education and innovation to break through ceilings that had existed only a generation earlier.

The early internet was built in coffee shops, university halls, and tiny offices filled with people who loved to debate, refine, and reinterpret ideas. Those rooms often resembled secular yeshivas with circles of brilliant, highly opinionated minds wrestling with abstractions late into the night. In that world, Jewish cultural presence was almost a gravitational force. Many founders, Jewish or not, found themselves influenced by the energy and intensity of that intellectual style.

The phenomenon extends beyond the founders of social media platforms to the cutting edge of artificial intelligence. Walk through the research labs at OpenAI, Anthropic, Google, or Microsoft and you will find Jewish researchers disproportionately represented, not because of any coordinated effort, but because the field rewards precisely the areas where Jewish thinkers have historically excelled—complex reasoning, abstract modeling, philosophical problem solving, and recursive analysis. In some ways, AI is simply Talmud with larger data sets.

Sam Altman: Jewish Icon of the AI Revolution

No account of Jewish presence in the modern tech landscape would be complete without Sam Altman, the most recognizable face of the AI revolution and, unintentionally, one of its most rabbinic. Altman is not an observant Jew in any traditional sense, and he does not present himself as a public Jewish figure, yet he embodies the very old Jewish archetype of the wandering philosopher-tinkerer whose ideas about human potential, and human danger, spill out into the world faster than anyone can process them.

Altman grew up in a secular Jewish household in St. Louis, the kind of environment where intellectual independence was encouraged and where curiosity was considered a virtue rather than a distraction. By the time he became the president of Y Combinator, a tech startup accelerator and venture capital firm, he was already a kind of secular talmudic figure in Silicon Valley, known for his ability to dissect complex systems, argue competing interpretations of technological risk, and imagine far reaching ethical dilemmas with the seriousness of an old world sage contemplating cosmic justice. His leadership at OpenAI pushed him even further into this quasi rabbinic role. As AI systems grew more powerful and more unpredictable, Altman found himself fielding questions that sounded less like engineering problems and more like halachic inquiries about the nature of creation, responsibility, and the future of humanity.

What sets Altman apart from earlier Jewish tech titans is the scope of what he's stewarding. Brin built a search engine, Zuckerberg built a social network, and Ellison built enterprise software. Altman is helping build a general purpose intelligence system that might one day make choices, generate ideas, and shape global knowledge infrastructures. That kind of project places him directly in the path of longstanding Jewish ethical concerns involving the role of the human creator, the dangers of idolizing one's own inventions, and the moral obligations that arise when tools begin to act with something resembling agency.

Altman's public presence has also made him a target for antisemitic conspiracy theories. As AI captured cultural imagination, far right and far left corners of the internet began spinning predictable narratives about Jewish control of artificial intelligence, recycling the old paranoia that has followed Jews across every technological revolution. The more powerful AI becomes, the more Altman's Jewishness becomes a projection surface for those fears, regardless of how he sees himself.

Yet Altman's influence also reveals something deeper about Jewish participation in technology. His work reflects a trait that runs through centuries of Jewish intellectual life, the willingness to wrestle with enormous, unanswerable questions like what is intelligence, what is freedom and what are the moral boundaries of creation, and how do we ensure that powerful systems serve human dignity rather than undermine it? These are not unfamiliar themes to Jewish thinkers, and they sit at the heart of Jewish ethical tradition.

So where does Sam Altman fit in? He stands precisely at the crossroads this chapter describes, a Jewish figure shaped by a long cultural lineage of questioning, reasoning, and moral anxiety, using those tools to guide a technology that could define the next era of human civilization. He did not ask to become the public philosopher of the machine age. But when the world began looking for someone to interpret the meaning of our new digital creation, it was a Jewish technologist, once again, who stepped onto the stage.

Dario Amodei: Jewish Ethics Meets AI

No account of Jewish presence in the modern AI landscape would be complete without Dario Amodei, the CEO of Anthropic and one of the clearest embodiments of the ethical anxiety at the heart of the machine age.

Before he became the head of Anthropic, Amodei had already established himself as one of the central architects of contemporary AI. He worked at Google Brain, later became Vice President of Research at OpenAI, and helped lead the development of major large language models. Anthropic itself was founded as an AI safety and research company devoted to building systems that are reliable, interpretable, and steerable. This framing reveals something essential about Amodei's role in this era: he is not merely trying to make intelligence more powerful, but to make power answerable to judgment.

That is what makes Amodei feel, in a deeply metaphorical sense, almost rabbinic. He keeps returning to questions that sound like ancient arguments about responsibility. *What happens when human beings create something whose capacities exceed their own? How do you govern a force that may become too complex to fully interpret? What obligations attach to the creator when the creation begins to act with something like autonomy?* Anthropic's public work is saturated with those questions, especially in its repeated insistence that frontier systems must be made understandable and controllable before society entrusts them with too much authority.

Like Altman, what separates Amodei from earlier Jewish-linked figures in technology is not simply his prominence, but the nature of what he is trying to steward. Previous generations built platforms, networks, marketplaces, and software empires. Amodei is helping build and govern systems that could influence research, labor, security, education, and human decision-making at civilizational scale. In his writing and Anthropic's policy materials, he has argued that highly powerful AI may emerge soon and that the central challenge is no longer whether these systems will matter, but whether human beings will remain morally serious enough to guide them. That places him squarely in the terrain of old Jewish concerns: the limits of human wisdom, the dangers of unrestrained power, and the demand that creation remain subordinate to human dignity rather than human vanity.

Jewish Futurism: Our Long Tradition of Imagining Tomorrow

If Jewish intellectual life is a centuries long bootcamp for innovation, then Jewish imagination has always been a rehearsal for the future. Long before Silicon Valley coined the term futurism, Jews were practicing it, sometimes in yeshivas, sometimes in literature, and sometimes around Shabbat tables where entire universes could be theorized between the soup and dessert.

Jews have been imagining tomorrow for thousands of years, not because it was fashionable, but because it was necessary. A people who survive through exile inevitably become futurists. Jewish futurism is not a niche literary genre, it is a civilizational reflex.

The Jewish Sci-Fi Brain: Asimov, Cyberpunk, and the Architecture of Tomorrow

It is no coincidence that some of the most influential writers shaping humanity's understanding of artificial intelligence were Jewish. The grandfather of AI ethics, Isaac Asimov, was a Russian born Jew who essentially invented the field of robot morality by writing, *"The Three Laws of Robotics"*, the most famous ethical code ever devised by someone not dealing with a regulatory agency. His laws were not just plot devices, they were Jewish ethical inquiry re-skinned for machines, and included the meaning of duty, the definition of harm, and the importance of being responsible for one's own creations. These are, in many ways halachic questions, legal thought experiments about behavior under conditions of uncertainty. Asimov was essentially writing The Three Commandments for robots.

Beyond Asimov, Jewish writers like Stanislaw Lem (philosophical AI satire before AI existed), Harlan Ellison (AIs that hate their creators), and Michael Chabon (Jewish alternate futures) all contributed to the conceptual landscape we now use to understand AI.

The Jewish Obsession With Utopia and Dystopia

Jewish futurism is not strictly optimistic or strictly bleak, but both. Jews have always oscillated between utopia (the messianic dream) and dystopia (every century of Jewish history). This oscillation produces a distinct speculative sensibility, one that holds that the world can be redeemed while also warning Jews to still keep a bag of supplies packed, just in case. Jewish utopia is always ethical and communal with focus on justice, dignity, freedom. Jewish dystopia is always historical, with expulsion, erasure, manipulation, and memories of loss. Both are present in modern Jewish contributions to tech, with the desire to build better worlds alongside an acute awareness that powerful systems can go sideways.

This tension is why Jewish sci-fi feels prophetic. Jews write of futures the same way they write history, with warning, imagination, and a touch of dark humor.

Why Jewish Futurism Matters for AI

The vast library of Jewish stories featuring Golems, Messiahs, heavenly scrolls, and prophetic visions, are, at their core, speculative narratives about the boundary between human and not human, and about what happens when creation exceeds its creator. AI is simply the latest version of this ancient question. In this way, Jewish futurism doesn't just predict the future, it interrogates it. It asks; *What kind of beings are we making? What moral obligations do we owe them? What moral obligations do they owe us? How do we prevent power from turning monstrous? How do we protect the vulnerable when systems grow too large?* Jewish futurism has always approached tomorrow with humility, fear, creativity, and stubborn hope, *the exact combination AI requires.*

The Sociological Logic of Jewish Overrepresentation

If you strip away the conspiracies with which the Jews are all too often charged, what you're left with as an explanation to their affinity for technology is sociology. Jewish culture prized literacy long before most societies did. By the Middle Ages, Jewish boys were required to learn to read while many Christian nobles couldn't. Diaspora life made Jews specialists in portable skills because when an expulsion edit was issued or national borders shifted, land was useless. Knowledge wasn't.

Moreover, the Jews are a communal people, and they settled in cities just as industrial scale production began. Cities were where innovation happened. Tight neighborhoods are where good human networks develop for mutual support. What antisemites see as conspiracy, is really just community.

Jews learned to innovate to survive. When their Temples were destroyed in the Babylonian and Roman conquests, they created a portable faith with a traveling God. Jews understand that stability is fragile and that creativity is mandatory.

Tech rewards exactly these patterns of curiosity, abstraction, mobility, analytical rigor, networked collaboration, and a high tolerance for uncertainty.

The Jewish Psychology of Innovation

Beneath all the historical, cultural, and sociological explanations lies something more intimate and more human. Jews do not innovate only because of circumstance or education or opportunity. They innovate because Jewish life, in its daily rhythms and emotional architecture, is a training ground for the very cognitive habits that fuel invention.

To be Jewish is to live inside a system where questioning is not merely allowed but expected. The first Jewish classroom is the Seder table, and the child's first assignment is "*Why is this night different from all other nights?*" Jewish children learn early that authority is provisional, that explanations are negotiable, and that every answer is only the starting point for another question. A mind raised in this environment does not freeze in the face of complexity, but leans forward.

Innovation requires comfort with ambiguity, hunger for understanding, the refusal to accept an answer at face value, and predisposition to ask if something could be different. Judaism has been cultivating those instincts for millennia.

Jewish life is also infused with the psychology of negotiation between the fixed and the fluid, the immutable and the pragmatic, and the obligatory and the improvised. Jewish law does not simply decree. It debates and it struggles with contradictions, exceptions, and edge cases. It allows for arguments that span generations, where a modern rabbi can respond to a medieval rabbi, who had responded to an ancient rabbi, as if the centuries between them were nothing more than a pause. To innovate is, in part, to move between what is known and what might be discovered. Jews learn this movement in childhood and acquire a mental flexibility unique to civilizations built around text and interpretation.

Then there is the psychological impact of diaspora. For two thousand years, Jews lived with the awareness that stability was temporary and reinvention was mandatory. Entire communities were displaced or uprooted themselves from Spain, France, England, Germany, Poland, Morocco, Syria, Iran, and others, only to rebuild their worlds from scratch in strange lands. Rather than a dramatic choice, reinvention was simply business as usual. Mobility, adaptation, multilingualism, community, and the ability to read unfamiliar environments were survival tools imposed on Jews over centuries of oppression. Those tools remain embedded in Jewish psychological DNA.

Jewish theology contributes its own layer. Unlike many religious traditions that emphasize certainty, Judaism places doubt at the center of spiritual life. Abraham questions God and Moses argues with God. The prophets criticize their own societies, often pointing the finger of blame for Jewish misfortune on Jewish corruption or unholiness. The Talmud is, essentially, a book length argument. This religious comfort with uncertainty mirrors the emotional reality of technological innovation, where breakthroughs emerge from the tolerance of confusion and the courage to explore the unknown.

Finally, there is an emotional dimension in the capacity to focus on the future. A Jew begins thinking of the future by considering the past. Jewish history, possibly because of its traumas, is relentlessly future-oriented. Every Jewish holiday, every prayer, every lifecycle ritual contains within it a vision of tomorrow—redemption, return, rebuilding, and healing. Jews have been imagining better futures for thousands of years, not as escapist fantasy but as a psychological necessity. That imaginative discipline is the beating heart of technological creativity.

Given all that, there is no mystery about the reasons Jews gravitate toward industries that rewrite reality, like computing, media, finance, science, and now artificial intelligence. Where else would a people trained for millennia in intellectual improvisation go but into the industry inventing the future?

The Human Truth Behind the Pattern

In the end, Jewish presence in innovation is not conspiratorial or somehow arranged by industrial puppeteers. It is the product of a people who survived perpetual displacement by mastering the only tool that could always be carried wherever they went, *the mind.*

As we turn toward the next chapter to answer the core question at the heart of this book, *Is AI good for the Jews?*, we carry this history with us. While it is clear that Jews have shaped the technological world, the real question now is whether Jewish values, instincts, and wisdom can help steer that world toward compassion rather than cruelty. Maybe the real Jewish influence on tech isn't control. Maybe it's conscience.

Part 6

Is AI Good for the Jews?

Chapter 27

So...Is AI Actually Good for the Jews?

A Jewish Conclusion to a Very Jewish Question.

There is no question more quintessentially Jewish than this one. That's not because the question is neurotic or paranoid, although, sure, there is a touch of that, but because it forces us to look directly at the intersection of power, vulnerability, responsibility, and change.

For five hundred years, every technological leap has presented the same double-edged promise. The printing press democratized Torah learning and standardized Jewish practice, but it also mass produced antisemitic pamphlets and woodcuts that shaped European hatred for centuries. Trains and steamships carried Jewish migrants to new worlds and also carried the Protocols of the Elders of Zion farther and faster than any rumor traveled before. Radio gave the world Jewish comedy and Father Coughlin, and film gave us Charlie Chaplin and the Nazis their most effective propaganda machine. The internet gave Jews community, visibility, education, and global connectivity, and also gave antisemites a global distribution pipeline.

If history teaches anything, it is that for the Jews, innovation always arrives like a new neighbor bearing both a freshly baked challah and a suspiciously heavy bag. For Jews, opportunity and danger, optimism and dread, liberation and vulnerability always seem to travel together. Every time new doors open, so do new trapdoors. AI is no different, so the question, then, remains: *Is AI good for the Jews?* The partial, unsatisfying, but honest answer is *it depends. It depends on us.*

If AI is trained on toxic data, it will reproduce toxic ideas, if it is trained on ignorance, it will produce ignorance, and if it is trained on centuries of antisemitic narratives, those narratives will gain a new vector—one that is sleek, scalable, and searchable. On the other hand, if the lessons of Jewish history prevail upon the creators of AI to train their agents on a worldview scrubbed clean of ideas that foster hatred and violence against Jews and others, AI can not only be good for the Jews, but great for humanity.

This is what the rabbis might say is the real question, *how do we make certain AI is good for humanity?* Judaism has grappled for millennia with questions about responsibility, interpretation, power, and moral ambiguity, which are exactly the questions AI forces society to confront, and so Jewish ethical wisdom is deeply relevant here. Concepts like tzelem Elohim (human dignity), lashon hara (speech ethics), pikuach nefesh (saving lives), and din rodef (ethical defense) offer frameworks that modern AI desperately needs.

If AI is going to be good for the Jews, it must be good for humanity. If AI harms vulnerable communities, it will harm Jews. If it amplifies hatred, it will amplify antisemitism. If it undermines truth, it will undermine collective memory. The Jewish lens on technology is ultimately a universal lens that calls on us to protect the vulnerable, honor dignity, preserve memory, elevate truth, and never outsource moral responsibility to machines.

A Closing Image

Imagine Jews gathering, online or offline, to learn, pray, teach, comfort, remember, disagree, and laugh, as they have done for millennia. Maybe the space is a synagogue, or maybe it's a WhatsApp group, or a Zoom room. The medium can change, but the Jewish conversation never will. The People of the Book have become the People of the Browser, and soon, perhaps, the People of the Algorithm. The challenge is to make sure that the algorithm does not replace the book, the screen does not replace the soul, and the machine does not replace the human encounter at the heart of Jewish life.

So, is AI good for the Jews?

Not yet. Not automatically. Not without effort. But it can be if Jews help shape it with the same creativity, vigilance, skepticism, compassion, and yes, chutzpah, that have carried them through every previous technological revolution. After all, if the universe is handing us a new creation story, the least we can do is add Jewish commentary.

Epilogue

A Blessing for the Algorithm

If you've made it to the end of this journey, mazel tov, you have officially completed the world's first comprehensive tour through half a millennium of Jewish technological anxiety, an experience for which there should probably be continuing education credits, or at the very least, a cookie.

In the tradition of our ancestors, who ended every difficult chapter with a blessing (or at least a discussion about what the blessing should be), it feels right to close this book with one as well. Jews have been blessing things for thousands of years. We bless bread, wine, fire, rainbows, mountains, new clothes, old friends, the moon, our children, and moments of survival that feel like miracles. So why not the algorithm? After all what is a blessing if not a way to show intention, an act of directing hope toward the world we want to live in. And intention might be the most powerful tool we have as we enter the next technological age.

A Blessing for the Algorithm

May the code we create always bend toward wisdom,
and may the systems we build remember the dignity of the humans who built them.

May our machines learn from the best of us,
and not the worst of us,

May they help us speak truth faster than lies, heal more than they harm,
and widen the circle of human understanding rather than shrink it so it is optimized for outrage.

May they preserve the fragile, stubborn, sacred stories,
that have carried so many of our communities, through fire and forgetting.

And may they remind us that intelligence, artificial or otherwise,
means nothing without conscience.

That even the most advanced machine cannot replace
the messy brilliance of human empathy, the courage of moral choice,
or the ancient, absurd, wonderful tradition of debating our way toward truth.

And, may we remember that every technology, no matter how powerful, is still just a mirror, and the responsibility for what it reflects rests with us.

And let us say, Amen.

About the Author

Craig Frank is a writer, entrepreneur, and global business advisor whose work sits at the intersection of Jewish history, emerging technologies, and the shifting media landscape. Drawing on decades of executive leadership, cross-cultural consulting, and lived experience in Israel, including service in the Israel Defense Forces, Craig brings a uniquely grounded and historically informed perspective to the challenges facing liberal democracies in the digital age.

Craig's career spans more than two decades of CEO-level leadership. As Partner and a key executive at Kaya Holdings, a publicly traded company, he oversaw corporate strategy, operations, compliance, and financial performance, building a vertically integrated platform with retail, cultivation, and proprietary brands. Before that, he co-founded The Tudog Group, a Tel Aviv–based international advisory firm that worked with more than 200 companies across 19 countries in the technology, healthcare, defense, and consumer sectors. That global experience, spanning the U.S., Europe, Latin America, and Israel. informs his understanding of how institutions adapt to disruption and how communities navigate uncertainty.

A keynote speaker at venues including the Florida Sterling Council, the International Project Management Association (Israel), Florida International University's Pico Center for Entrepreneurship, and the Israel Center for Entrepreneurial Studies, Craig has lectured across the Americas and beyond. He has published more than 300 articles on business and strategy, edited two works of Israeli military memoir ("A Soldier's Story" by Rafael "Raful" Eitan and "A Warrior's Way" by Avigdor Kahalani), and is the recipient of the President's Volunteer Service Award and the New York Attorney General's 3 C's Award for Character, Courage, and Community Concern.

He holds a Bachelor of Arts in Political Science and Sociology from Queens College, City University of New York, and completed graduate studies in International Relations at the CUNY Graduate Center.

www.ingramcontent.com/pod-product-compliance
Ingram Content Group UK Ltd.
Pitfield, Milton Keynes, MK11 3LW, UK
UKHW040232290926
15471UKWH00002B/152

9 781968 919